Saint Peter's University Library
Withdrawn

Saint Peter's University Library
Withdrawn

VDT Health and Safety

VDT Health and Safety

Issues and Solutions

Elizabeth A. Scalet

T.F.M. Stewart
Consulting Editor

Kate McGee
Research Associate

Ergosyst Associates
Lawrence, Kansas ● London

©1987 by Ergosyst Associates, Inc.

All rights reserved. No part of this publication may be reproduced, stored in a retrieval system, or transmitted in any form or by any means, electronic, mechanical, photocopying, recording, or otherwise, without written permission of the publisher.

Printed in the United States of America by W.A. Krueger Co.

Published by Ergosyst Associates, Inc., 910 Massachusetts St., Suite 602, Lawrence, Kansas 66044-2975 U.S.A.

Elizabeth A. Scalet
T.F.M. Stewart, Consulting Editor
Kate McGee, Research Associate

Design and production by Publication Management Services

Library of Congress Cataloging-in-Publication Data

Scalet, Elizabeth A., 1948-
  VDT health and safety.
  Includes bibliographies and index.
  1. Video display terminals—Hygienic aspects.
I. Stewart, T.F.M. II. Title. (DNLM: 1. Computer Systems. 2. Occupational Diseases—prevention & control.
3. Radiation Effects. WA 470 S281v)
RC965.V53S27   1987   363.1'89   87-12756
ISBN 0-916313-13-1

RC
965
.V53
.S27
1987

The symbol of the electronic information revolution is the flickering Cathode-Ray Tube (CRT) screen . . . An observer from another planet might wonder what new religion arose on earth whose converts sit hours a day in solitary worship before tens of millions of blue-eyed oracles.

Arnold S. Wasserman
"Office Automation:
Post-Industrial Revolution,"
Ergonomic Considerations of
Visual Display Units

Contents

Illustrations

Figure

Preface

T.F.M. Stewart

When I was invited by Ergosyst Associates to be consulting editor of *VDT Health and Safety*, I was pleased, and flattered, to be asked to help with this important initiative. I have long been concerned that the widely available sources of information about VDT health issues are often flawed.

Some are flawed in the sense that they are factually wrong. Many journalists seem to be entirely happy to write about VDT health issues without understanding the topic at all. Sometimes the errors are subtle or only slightly misleading. Sometimes there are misunderstandings which get passed from article to article and are never rescinded.

Others are flawed by being biased towards a particular point of view. It is very easy to give the same information a quite different ring by juxtaposition or by changing the emphasis. For example "researchers fail to exonerate VDT from abortion scare" may refer to exactly the same report as "researchers find no link between VDT and pregnancy outcome." But they mean very different things to the pregnant VDT operator.

I was therefore pleased that Ergosyst, with its track record of objective and informed comment, was going to publish a source book on VDT health and safety.

However, I was also a little apprehensive. I was concerned that it might prove impossible to explain the complexity and uncertainties of the VDT health debate in an interesting and readable way. The nature of the controversy is such that it defies simple, truthful explanations. True explanations are unlikely to be simple and simple ones are unlikely to be true.

But my fears were unfounded. The Ergosyst team, especially Beth Scalet, has done an outstanding job in bringing together a wide variety of original information and has distilled it into a form which is both useful and usable.

Their job has not been made any easier by the noise and confusion which surrounds VDT health issues. As I have said elsewhere, VDT health issues have been researched and debated for a number of years, but regrettably, not always in that order. Much of the noise has been caused by our incomplete understanding of what is going on when someone uses a VDT. The growth of sound research around the world has helped to provide some answers, and distinguish the myth from the fact.

But it is still not as straightforward as might be supposed. Although the VDT is just a piece of equipment, it is only one component of a wider system or series of systems. These systems include the computer system itself, the business system and procedures it is used to support, and the organizational system to which the user belongs. VDT health problems may occur in any or all of these systems.

Indeed, it is often the interactions between these systems which are most problematic. For example, a VDT which can be used quite acceptably for checking system statistics in the controlled environment of a computer room may be totally unacceptable for extensive proofreading in a brightly lit office. It is the interaction between the VDT, its environment, and how it is used that causes the problem. In other words, some of the health problems are VDT specific, some are VDT work related, and some are more to do with the nature of modern office work.

But it is not always clear what we mean by a "health problem." To many people, a health problem is something which causes long-lasting or permanent damage. When we talk about good health we often mean little more than the absence of injury or illness. But health is a much more positive concept. It has to do with well-being. Excessive fatigue and discomfort may not result in permanent damage or illness but they are still health problems.

However, those who wish to exonerate VDTs ask the question "Is using a VDT any worse for your eyes than watching TV, or is it any worse for your posture than using a typewriter?" The answer to this (after several years of research) is probably "No" (at least in most situations). Using a VDT is no worse for your health than many of the things that people have tolerated for years.

The VDT industry breathes a sigh of relief. But what about the users? Is it reassuring to know that you are only damaging your back by sitting badly at a VDT as much as you damage it at a normal desk? Why put up with it, if you do not have to? If we can avoid the risk, however small, shouldn't we do just that?

In the real world, it is not that simple. Why should those who supply VDTs be blamed for all society's ills? We need some kind of perspective and balance. That is where this report comes in. It addresses the issues head on. It tries to be balanced and present the facts where these are known. It cannot be unbiased. That is not really possible but our bias is towards those who want to make use of the VDT (as we all do ourselves) and also avoid unnecessary risks to health.

The history of the VDT health debate is one of uncertainty and confusion. Various parties have been guilty of hype and extrapolating far beyond their data to make sweeping pronouncements. Lay people have been astonished at the disarray in science and much good science has been discredited as a result. Yet, this topic is probably no worse than many other scientific debates. It just happens to be being conducted in the full glare of publicity and many people are passionately interested in the outcome.

But science is always about uncertainty. Research is there to discover the unknown. If there were no surprises, there would be no point in research. Every scientific finding carries with it the risk that it was a chance finding and not a true reflection of the world. No single result should be enough to convince us to make dramatic changes to our world. They need to be replicated and corroborated. They need to be supported by an explanatory framework which is believable before we rely on them fully.

Now there are times when one dramatic discovery should really make us sit up and take notice, for example, if we believe that something poses a risk to health. But where complex interactions are involved, we must be certain that we identify the real cause. What we substitute for the risky activity must not make matters worse. The desire for urgent action should be tempered by the realization that knowing about the problem does not make it any worse. If harm is being done, it is not any worse for the victims than it was before we made our discovery. We have as much time as we had before to take action, we are just more aware of what is really happening.

However, we can make matters worse by encouraging a change which is wrong or by causing unnecessary alarm and concern. For example, some have argued that pregnant operators should be transferred from VDT work automatically. In some organizations this might involve changing to work which requires the operator to stand all day, which is actually riskier to the fetus. And even just worrying about it may be worse. New evidence should be added to our total stock of knowledge and not overreacted to blindly. Lead is a dangerous poison yet some people have argued that aprons containing this known poison should be worn by pregnant women to protect them against the remote possibility of some unknown VDT radiation. That is not the action of an informed person taking a balanced view of risks and making a rational decision about uncertainty, risk, and life.

So how do we minimize risks? One argument is that if it does not cost much to protect against a risk, no matter how slight, then it has to be worthwhile. Certainly there are areas in which that is true and it probably does make sense to design new VDTs to minimize their emissions of all kinds provided it is not too difficult or expensive. However, sometimes there are side effects and these can be difficult to quantify. For example, some electromagnetic shielding reduces image quality and resolution. Is it worth it? Probably not. Also, given the history of the debate, any such action to shield VDTs now may be exploited as a sign that there must be some more sinister evidence to make the manufacturers do it. This is a "Catch 22" situation. If the manufacturer refuses to do the simple change because he believes that it is unjustified then he is accused of irresponsibility. If he does it because people ask for it, he is accused of supporting the view that there really is something to hide after all. This is an example of how the debate has polarized the issue and made it all the more difficult for reasonable people to make reasonable decisions.

What I would like to see is more people taking the initiative for their own future and not relying on experts all the time. No one is more concerned about your own well-being than you are yourself.

So, drawing this all together, what can we conclude? One question that I am often asked which brings the debate sharply into focus is "What do you tell a pregnant VDT operator?" I take that responsibility very seriously.

The first thing that I say is do not worry about radiation, adding to your worries with radiation scares is certainly not good for you. What you should do is look after yourself. If you avoid excessive discomfort and fatigue, carrying on your normal work at a VDT is probably better for you than any major and disruptive change to your working life. But if you find it difficult, and many people do at some time in their pregnancy no matter what type of work they do, a reasonable employer will be sympathetic and understanding about helping you to modify your work to make it easier.

But really, this is good advice for everyone, not just VDT operators. Everyone should take care of their eyes, their musculoskeletal system, their posture, and stress levels.

Indeed, one lesson that has emerged most clearly in the VDT health debate is that it never does any harm to improve the ergonomics of the workplace. It may solve the health problem completely. Even if it does not, it is probably good for productivity. It is certainly good for morale and that helps staff and management.

However, improving the ergonomics requires sound judgment. Sound judgment needs information and that is why I commend this source book to you. It provides the individual and the organization with the information to help them to make informed and sensible choices about health, comfort, and personal productivity. Use the information wisely.

1 Introduction

- Between fifteen and twenty million workers use VDTs daily, and it is anticipated that by 1997, 50 percent of American workers will use them.

- Complaints about eyestrain, backache, and other aches and pains among VDT operators are widespread and real.

- Allegations that VDT use may be harmful to human fetuses have led to many studies. At this time, no link has been proven, and no strong correlation is likely to be discovered.

- Recent investigations suggest that job stress is one of the most important factors contributing to VDT operator complaints.

- Almost all of these complaints can be alleviated or avoided by proper attention to the workstation, the work environment, and the design of the work.

Between ten and fifteen million computer terminals, also called VDTs (for visual, or video, display terminal), are in use in the United States today.[1] Between 1972 and 1980, the number of operators of computer and peripheral equipment increased by 170 percent.[2] Between fifteen and twenty million workers use VDTs daily, and by 1990 the number of VDTs in use is expected to increase to thirty-eight million. By 1997, 50 percent of Americans will be using VDTs at work.[3]

Most of this change has occurred rapidly, within the last decade. This is due to the development of the microchip, which has permitted manufacturers to build smaller, less expensive, more powerful computers.

When any new technology is introduced rapidly into an existing system, unexpected changes are likely to occur. The introduction of the automobile led to unforeseen needs for such things as traffic lights and freeway systems, and it took many years to develop a coherent "systems approach" to traffic planning and safety. Likewise, the technology itself became more refined, producing faster, more comfortable autos, with more and more features. If retired U.S. Navy Rear Admiral Grace Hopper (who was instrumental in developing the computer language COBOL) is correct in saying that the computer revolution has just begun, and that compared to what is on the horizon today's computers are like "Model T's,"[4] we can expect the computer's development to continue to produce change in our workplace systems, change that we must be prepared for if we are to manage it.

The VDT is already changing the workplace, and indeed, the nature of office work.

Overview of Health Concerns

VDTs are not just part of a changing workplace. They are also one product of a burgeoning industry, and as such they are subject to the same questions that any new mass market product must face. Is it safe to use? How do we know? Can it be made safer? How much will it cost to make it safer? Is it going to put people out of work? Will workers overcome their initial fears and adapt to the new technology?

In the case of VDTs, the pressure from management to automate, to "keep up with the Joneses" and be competitive, has been met on the other side by pressure from workers and their representatives, pressure fueled by a number of health complaints. This is not especially surprising. Specialized occupations have always presented unique demands on the human body. Something as simple as a carpenter hitting his thumb with a hammer is, in a real sense, an occupational hazard. Recent attention to more catastrophic (and therefore sensational) occupational hazards such as black lung disease and asbestosis has made it clear that the adverse health effects of many workplace hazards do not become apparent immediately, but may take years to develop.[5] It is in this social climate that the VDT has been introduced, and so it is no wonder that many users greet the arrival of the display terminal with suspicion and anxiety.

According to Leela Damodaran, a specialist in the field of ergonomics (the study of the relationship between humans and their work), "Fears about new technology may be expressed as health hazards because these constitute the only legitimate reasons in our society to reject the technology."[6] These fears are expressed in numerous ways. Workers may feel that their competence, their control, their power, or their job security is threatened by new technology. They may simply fear the unknown. Such concerns may be quite stressful for the worker.[7]

In addition to stress, other complaints among VDT users have surfaced. Backaches, eyestrain, burning or itchy eyes, headaches, neckaches, fatigue, moodiness, and nausea have become common enough among terminal users that a new catchphrase is emerging: VODS—Video Operator's Distress Syndrome.[8] Facial rashes among operators have also been reported, especially in Scandinavia.[9]

More serious allegations have also been made. A worker in New York state was awarded workers' compensation for cataracts which she claimed resulted from the radiation from her VDT. That decision has since been set aside, however, and the case scheduled for rehearing.[10] Australia has reported a high incidence—20 to 30 percent of the full-time keyboard users— of wrist and hand injuries, some of which have required surgical repair or have resulted in partial or complete manual disability.[11] Most disturbing of all are reports of clusters (unusual groups) of reproductive problems (including spontaneous abortions, stillbirths, and birth defects) among VDT operators at worksites in the United States, Canada, and elsewhere.[12]

Throughout this first decade or two of VDT use, researchers have conducted field studies, laboratory experiments, and epidemiological studies to investigate these complaints. In some areas there is still disagreement. In others, there is a great deal of unanimity. In still others, two entrenched camps may oppose each other more because of political differences than scientific ones. In answer to the question "Are VDTs a health hazard?" it might be hard to get a definitive "yes" or "no" from most sources, but it is now likely that it would not be hard to get a "probably not" from many of them.

The growing consensus about many VDT health issues was apparent at the first International Scientific Conference: Work with Display Units, held in Stockholm, Sweden, in the spring of 1986. Over 300 papers were presented by representatives from virtually every sphere—labor, government, industry, science, and management. This landmark conference allowed the exchange of the most recent knowledge available about VDTs and health, and while it included some controversial research findings in the area of extremely low frequency electromagnetic field effects, it also demonstrated the advanced state of understanding that has developed in such areas as visual complaints and musculoskeletal strain.

This does not mean that there are no problems. Complaints of discomfort, both visual and postural, are certainly real, although most cases involve short-term discomfort and not permanent injury. The concerns of VDT operators regarding reports of reproductive problems are also disturbing and anxiety-producing, even though no link between such problems and VDT use has been established nor is likely to be.

However, in light of the research that has been reviewed for this report, it seems that the problems of visual and postural discomfort which VDT operators experience are not related to any intrinsic property of VDTs, but rather to a lack of understanding about how to put VDTs into the office setting in ways that will allow people to work comfortably and efficiently.

Reports of cataracts caused by VDT work have been convincingly refuted, and no permanent visual damage due to VDT work has been demonstrated, although eyestrain is common. Similarly, the muscular complaints that have been reported are most often short-term effects which can be relieved by rest and by attending to the office environment. In most cases where injury, not just discomfort, has been reported (such as repetition strain injuries to the wrist or hand) it seems very likely that improved workstation design, more frequent rest breaks, and better understanding of job design could have averted the problems.

For several reasons, the reports of clusters of abnormal pregnancy outcomes are more complicated to evaluate. The issue itself is emotionally provocative, and if any hazard were discovered in such a pervasive new technology its public health impact would be enormous. Thus this issue has received a great deal of study in spite of the fact that the statistical risk— if it were shown to be true— would be quite small. The clusters do not by themselves mean that any particular entity caused the events. According to complicated probability measures, they are very likely random groupings which have occurred by chance.

Investigations of this question have been more difficult than those of other health allegations surrounding VDTs, because it is not possible to study this question through direct experimentation. Results from animal studies are hard to translate to humans, and observational studies of human populations (epidemiological studies) are time-consuming, costly, hard to design, and hard to evaluate, especially when the question being studied is an event (such as birth defects) which is known to occur very infrequently.

One of the properties of VDTs which has been most strongly considered as a possible cause of reproductive harm is the radiation emitted by VDTs. Suspicion has narrowed to a particular type of low-level radiation which until recently had not been thought to have any effect on humans. In the last few years, however, scientists studying this radiation for other reasons have learned new and surprising things about how it may affect people. As a result, VDTs are rather accidentally in the middle of a highly volatile scientific debate about newly discovered properties of certain kinds of radiation. The very existence of this debate has probably prolonged the uncertainties about the effects of VDT radiation on pregnancy, even though the level of this radiation emitted by VDTs is quite low in comparison to power lines and many common household devices.

More experiments are being conducted in this area, but at present it seems likely that any connection between VDTs and reproductive failure is so small that it cannot be reliably measured. Certainly it is less than that of other known risk factors such as smoking and drinking. Should any connection between VDT work and reproductive problems be found it is more likely to be the result of work-related stress. This stress may be exacerbated by the ways in which VDT work is designed, but it is not due to any characteristic of the VDT itself. So VDTs are probably safe. But because the potential damage is so serious, and because the impact could be so profound, this issue will continue to be investigated for as long as public concern is strong enough to justify the time and expense involved.

While this issue is being resolved, however, there is much that can be done to improve the conditions of VDT work. In the chapters which follow, the various complaints which are common among VDT operators will be examined, and the causes of those complaints (as they are currently understood) will be discussed. In many cases, there are specific suggestions about what can be done to improve the situation. But the most important "finding" of this report may be the appreciation of the human system as a complex, highly sensitive mechanism which deserves at least as much consideration in its use at work as the highly complicated and sensitive computer systems that are the tools of today's office worker.

Further Readings

Çakir, A.; Hart, D.J.; and Stewart, T.F.M. *Visual Display Terminals: A Manual Covering Ergonomics, Workplace Design, Health and Safety, Task Organization.* John Wiley. 1980.
   One of the earliest and still one of the most comprehensive examinations of VDTs in the workplace, this manual presents detailed discussions of design recommendations for VDTs, workstations and chairs, and lighting, as well as guidelines for other environmental variables and for task design. Appendices include an ergonomic checklist, a proposed eye test for VDT operators, a bibliography, and an extensive glossary.

DeMatteo, Bob. *Terminal Shock.* NC Press Limited. 1985.
   This is a forceful presentation of operator concerns by a Canadian labor representative. The extensive discussion of radiation concerns ably demonstrates the activist position which labor has adopted on this issue. Appendices include a questionnaire for individual operators about many aspects of VDT work, recommended tests for radiation effects, recommended eye exams, and a summary of legislative action.

Sauter, Steven L.; Chapman, L. John; and Knutson, Sheri J. *Improving VDT Work: Causes and Control of Health Concerns in VDT Use*. The Report Store. 1986.
This concise practical guide for operators discusses common complaints, explains the physiological background and probable causes, and provides down-to-earth solutions which the operator can undertake him/herself. Appendices include checklists for tables and workstations, exercises to relieve muscular and visual discomfort, a relaxation technique, references, and bibliography.

Tuthill, Robert W. *Assessment of the Literature on Health Effects Related to Video Display Terminals*. Final Report to the Massachusetts Department of Health. State of Massachusetts. 1985.
This well-balanced and thorough review of the literature pertaining to VDTs (through early 1985) includes complete bibliographic details along with the author's conclusions and recommendations.

2   VDTs and Radiation

- VDTs, like many other things, produce some radiation.

- Across the range of what is called the electromagnetic spectrum there are many different types of radiation, from the damaging gamma, beta, and x-rays to visible light, radiowaves, and power frequencies which are regarded as relatively safe.

- Repeated measurements of emissions from VDTs have never shown any appreciable x-ray emissions; often these emissions are so low they don't register on sensitive instruments.

- Because x-ray emissions are so low, they have been ruled out as the cause of any reported adverse pregnancy outcomes among VDT operators.

- Reports that emissions from VDTs may cause cataracts have been convincingly refuted.

- The results of research into the effects of other radiation from VDTs, the very low and extremely low frequency pulsed fields, have been inconclusive so far.

- It seems likely that radiation from VDTs will not be found to play a significant role in operator health concerns. Other factors such as stress are more likely to be important.

- Meanwhile, pending further results, some governments are proceeding with caution by requiring better shielding from these fields.

Many common objects, such as limestone, wood, and even people, emit radiation. The most common form of radiation is visible light (optical radiation). VDTs also emit certain types of radiation in the course of normal operation. There is no debate over this. The very purpose of a display terminal—the presentation of information via illuminated characters—means that light, which is optical radiation, is intentionally emitted. The question is whether VDTs emit radiation of any type at levels sufficient to be harmful to humans or human fetuses.

Radiation is simply any form of energy that moves in waves or particles, even in a vacuum. But the word *radiation* has taken on a primarily negative connotation since the discovery of the harmful effects caused by certain types of rays such as x-rays and gamma rays. More recently, connections between overexposure to certain types of radiation present in sunlight and the development of skin cancer have received a great deal of attention. Radiation cannot be tasted, heard, felt, or sensed (although heating produced by high exposures may be felt after the fact, and a few individuals may be able to hear microwaves). Only one kind of radiation, visible light, can normally be seen. As a result, the nature of radiation is elusive, its effects are known to be powerful, and its presence is difficult to detect. Thus, discussions of radiation of any kind can create anxiety.

In the case of radiation emissions from VDTs, this anxiety has been fueled by reports of cataracts and reproductive failure among VDT operators. Research has never proved any connection between these events and VDT use, and radiation emission measurements have repeatedly been far below any levels known to cause health effects. Nonetheless, the serious nature of the allegations and the elusive nature of radiation have kept the controversy alive.

Radiation and Its Health Effects

In order to understand the emissions from VDTs and their possible effects, it is useful to have a brief introduction to radiation in general and then to examine each kind of radiation emission more closely.

Electromagnetic radiation is always in the form of a wave composed of two parts: an electric field and a magnetic field. The energy of radiation is measured according to its frequency (the number of waves that pass a given point in a specific time period, usually one second), which is expressed in Hertz (cycles per second, abbreviated Hz). The higher the frequency, the more powerful the radiation. Frequency is inversely related to wavelength (the distance from point to point between waves); the higher the frequency, the shorter the wavelength.[1] The various types of radiation are distributed across what is called the *electromagnetic spectrum*, which is itself usually divided into two basic categories, ionizing radiation and non-ionizing radiation (see fig. 1).

Radiation with the most energy (the highest frequencies) is called ionizing radiation. The various types of ionizing radiation have short wavelengths compared to the less active radiations found in the non-ionizing portion of the spectrum. As the wavelengths become very long, as they do in the lowest frequencies of the non-ionizing range, the two parts of the wave — the electric field and the magnetic field—become more and more separable into distinct components. At these frequencies, researchers must take separate measurements of each field in order to get an accurate representation of radiation levels. In the discussion below, each segment of the electromagnetic spectrum will be discussed separately, first in terms of its general characteristics and effects, and then in terms of the amount emitted by VDTs and the potential effects of exposure to such levels.

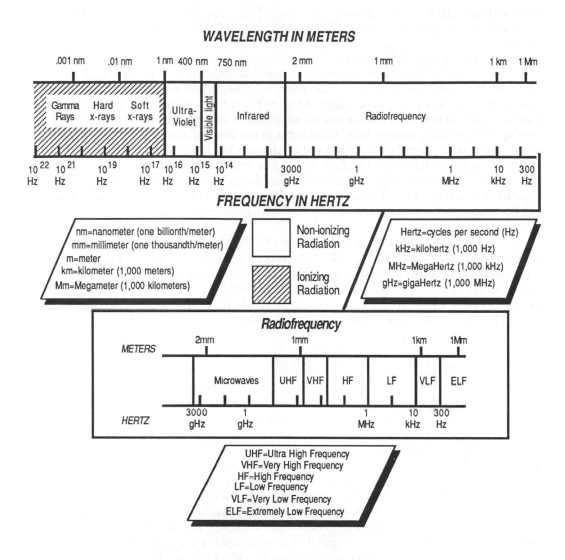

Fig. 1. Electromagnetic Spectrum with Enlargement of the Radiofrequency Range

Ionizing Radiation

Ionizing radiation includes alpha and beta particles, gamma rays, hard x-rays, and soft x-rays. Alpha and beta particles and gamma rays are produced by radioactive decay of such isotopes as Uranium 238. X-rays are produced when high-speed electrons are slowed down or stopped. When this happens, these electrons give up their energy in the form of x-radiation. Soft x-rays (from 0.1 to 1.0 nanometers, or nm, in wavelength) are less penetrating than the higher frequency hard x-rays (from 0.001 to 0.1 nm in wavelength).[2]

Ionizing radiation is known to cause cellular changes. It affects living matter by breaking chemical bonds and charging (ionizing) neutral molecules.

This may cause immediate cell death, a reduction of cell function and subsequent failure of body function, or cell damage, which increases the probability that the cell will reproduce itself in unpredictable ways. Unpredictable cell reproduction can produce mutations which may lead to radiation-induced cancers or damage to genetic information. Low dosages of ionizing radiation may produce minor reversible blood changes, but at higher dosages the symptoms progress from nausea, vomiting, and reduced production of white blood cells, to diarrhea, loss of appetite, and malaise, and finally to loss of hair and hemorrhaging followed by death.[3]

The effects of ionizing radiation have been studied intensively. This research resulted initially from the effects of ionizing radiation from the atomic bomb on the populations of Hiroshima and Nagasaki, and later from investigations of the now-recognized effects of the uncontrolled medical use of soft x-rays in treating a wide range of disorders such as acne and adenoids (common practice in the 1950s) and the use of radioactive materials like radium in watchmaking.

Over the last four decades our understanding of the biological effects of this type of radiation has been carefully refined. Standards for exposure to ionizing radiation have been developed based on this body of research. These standards have led to much stricter control of such common exposures as medical and dental x-rays. However, even these standards may be questioned since some experts feel there is no "safe" level of exposure and have adopted the ALARA (As Low As Reasonably Attainable) principle.

Because the harmful effects of ionizing radiation are well known, especially its ability to harm embryos and alter the genetic information contained in human

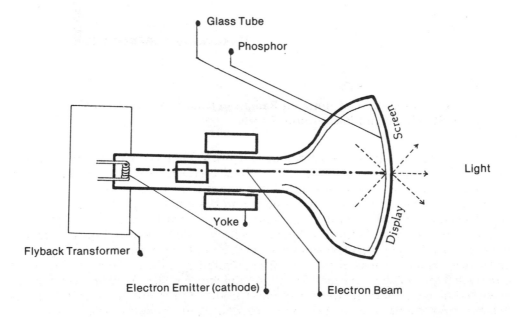

Fig. 2. Cathode-Ray Tube

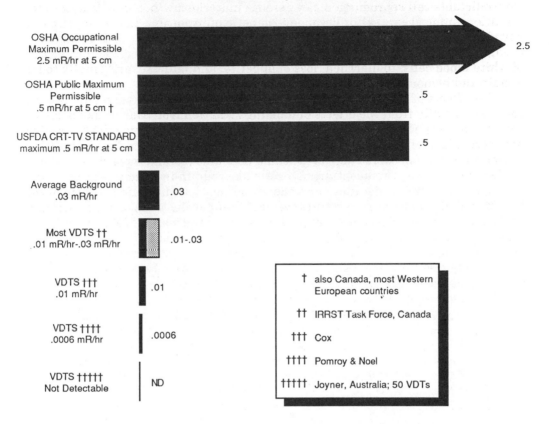

Fig. 3. Emissions of Ionizing Radiation from VDTs Compared to Standards and Background Levels

chromosomes, emissions of such radiation from VDTs have been a topic of great interest. In particular, whenever adverse health effects are reported among VDT operators it has been considered important to investigate any potential relationship between radiation emissions and the reported problems.

VDTs do not produce or emit gamma, alpha, or hard x-radiation. The cathode ray tube VDT (which is the most common, and the only type considered here) produces images on the screen as a result of the activation of phosphors on the screen face. These phosphors are activated by a beam from an electron gun in the back of the VDT (see fig. 2). As the fast-moving electrons are slowed down by the phosphors at the screen face (producing the image), small amounts of soft x-rays are produced, but only an insignificant amount of it escapes. Almost all of the x-rays are contained by the coated glass face of the VDT. Every reliable study has shown emissions of ionizing radiation from VDTs available in the marketplace to be below the level of background radiation, the level naturally occurring in the environment from sources such as sunlight and radiant objects (see fig. 3). Frequently the level of ionizing radiation from VDTs is so low that it cannot even be measured. In addition, sample models of electronic devices—including VDTs—are regularly tested by the U.S. Food and Drug Administration (USFDA) for emissions. Inadequately shielded or otherwise defective models are not permitted into the marketplace.[4]

In spite of these findings, some workers and labor groups continue to express concern that ionizing radiation from VDTs might be harmful to operators or to developing fetuses and have recommended that pregnant operators in particular be protected from such radiation. Some have proposed that pregnant workers wear lead aprons to shield themselves from radiation. While the issue of reproductive harm from VDT use is still being debated, this aspect of the controversy can be laid to rest: ionizing radiation from VDTs does not reach levels sufficient to be implicated in adverse pregnancy outcomes. Lead aprons do not shield the operator from some of the other emissions which have been of concern. Thus, even if other emissions produced by VDTs are found to be harmful, there is no reason to use such devices. In addition, these aprons are heavy and can lead to postural and thermal stress which are in themselves potentially harmful for the pregnant woman and her unborn baby.[5]

Non-ionizing Radiation

Non-ionizing radiation does not have sufficient energy to strip electrons away from molecules or break chemical bonds, but it can cause the rotation or vibration of molecules. This energy is then spent as heat (this is how microwave ovens cook). The non-ionizing range includes ultraviolet radiation; visible light; infrared radiation; and radiofrequency radiation, a broad range of frequencies which includes microwaves, communications frequencies, very low frequencies (VLF), extremely low frequencies (ELF), and power frequencies. Some sources treat microwave radiation as a separate range in itself.

Ultraviolet (UV). Ultraviolet radiation (from 180 nm to 400 nm) is present in sunlight. The higher frequencies (above 200 nm) are mostly absorbed by the ozone in the atmosphere. Lower frequencies are well absorbed by window glass. UV causes skin burning and triggers the body's protective tanning reaction. Chronic exposures or repeated extreme exposures may also increase the likelihood of skin cancer. Exposures to the eye above the Threshold Limit Value (TLV) can cause conjunctivitis (inflammation of the lining of the eye), keratitis (inflammation of the cornea), pain, light intolerance, and cataracts. The most common source of UV radiation is sunlight. UV is also present in industrial applications such as electric welding lamps and germicidal lamps.[6]

Some VDTs produce small amounts of UV radiation, but most of it is absorbed by the glass screen. While high levels of UV may have detrimental effects such as sunburn and, in sufficient exposures, cataracts, no international standard or recognized body of scientific work has shown biological effects for the levels of ultraviolet which are emitted from VDTs. In fact, emissions of UV are at levels less than one-one-thousandth of the safety standard.[7]

Optical radiation (visible light). Optical radiation is in the range of 400 nm to 750 nm. The energy spent by optical radiation is in the form of visible light. At most exposures visible light is not harmful. Sunlight, while it is composed primarily of radiation in the visible light range, also contains some UV radiation, along with smaller amounts of cosmic rays, which include some ionizing radiation.

Optical radiation is the primary emission from VDTs. It is the result of the activation of phosphors on the screen face by beams from the electron gun in the back of the VDT. Even so, the emission levels are less than one-one-thousandth of the National Institute for Occupational Safety and Health (NIOSH) standard, well below levels which could be harmful.[8]

Infrared radiation. Infrared radiation (from 750 nm to 0.1 centimeter) is simply what we would normally call radiant heat. It can cause skin burns and injury to the cornea, iris, retina, and lens of the eye. These effects are due to heating, and for most extreme infrared exposures the sensation of pain will alert the body to danger almost immediately. Some types of chronic high-level exposures to infrared radiation are known to have adverse effects. The best known is heat cataract, also called glassblower's cataract. Common sources of infrared include commercial heat sources and processes like welding or heating of metals or glass.[9] Some VDTs emit very small amounts of infrared radiation, at levels around one-one-thousandth of the standard.[10]

The next area of the electromagnetic spectrum is actually a group of frequencies. While these frequencies are often discussed together as *radiofrequency radiation*, the differences among these frequencies are of particular interest with regard to VDTs, and so each group will be discussed here individually. The radiofrequency range of the electromagnetic spectrum is a subset of the non-ionizing range, and includes microwave radiation, communications (radio, TV) frequencies, very low frequency (VLF) radiation, and extremely low frequency (ELF) radiation. At the extreme lower end of this range are the power frequencies, which carry electric current.

Microwave radiation. The microwave frequencies (from 300 gigaHertz [gHz] to 30 megaHertz [MHz]) include SHF (super high frequency) and EHF (extra high frequency) bands. Satellite communications, airborne weather radar, and shipborne navigational radar employ SHF. SHF is used in microwave point-to-point transmissions. EHF is used in radio astronomy and cloud detection research applications. At high exposures, microwave radiation can produce thermal effects such as heating of the eye, which may cause cataracts, or heating of the testes, which can damage sperm cells.[11]

In the 1960s an incident occurred which led to intensive examination of biological effects of low-level exposure to microwave radiation. It was discovered that the Soviets were bombarding the U.S. embassy in Moscow with microwaves at levels well above permissible Soviet population exposure levels, but below those permitted in the U.S. for occupational exposure. While it was at first thought that the Soviets might have been trying to interfere with electronic transmissions, this was later ruled out. Investigators then considered the possibility that the intention was to harm U.S. personnel or affect their behavior in some way. The actual intent, if known, has not been revealed. Some unusual health problems were reported to have occurred among embassy personnel, both at the time of service and after leaving service in Moscow. Among the alleged health effects were breast cancers and blood disorders, including leukemia, along with behavioral disorders such as depression. However, no causal links to microwave exposure have been demonstrated and the report of an investigation of the health status of embassy personnel conducted at Johns Hopkins University found no consistent pattern of unusual health conditions.[12]

When the first reports of cataracts among VDT users were publicized in 1977, ionizing radiation emissions were suspected, since it has been established that ionizing radiation can cause cataracts. However, VDT emissions were so low that ionizing radiation was ruled out as a possible cause in these cases. Suspicion then fell on possible microwave emissions from VDTs, since microwaves are known to cause cataracts at sufficiently high exposure levels. However, after extensive testing on many brands and models of VDTs, it is clear that there are no measurable microwave emissions from VDTs, and that they therefore do not represent a hazard for the VDT user.[13] In most models, there is not even a mechanism by which microwaves could be produced. It is possible that VDTs of a different design might include a mechanism which could produce microwaves. Therefore continuing measurements should be made by regulatory agencies such as the USFDA. However, at this time, microwaves cannot be considered to be in any way implicated in VDT health concerns.

Communications frequencies. The logic circuits of the VDT generate some radiation in the high radio frequency range, mostly UHF (ultra high frequency) and VHF (very high frequency). VHF is commonly used for radio and TV transmission, air traffic control, and radio navigation. UHF is emitted in TV transmissions, citizens' band radio, and meteorological radar. Levels of UHF and VHF emitted by VDTs are reported at no more than two-one-thousandths of the safety limit, and some reports claim these emissions are no more than one-ten-thousandth of the safety level.[14]

Very Low Frequency (VLF) and Extremely Low Frequency (ELF) electromagnetic radiation. The greatest amounts of radiation produced by VDTs, next to visible light, are VLF and ELF electromagnetic radiation. These are emitted from the line output transformer, also called the flyback transformer.[15] The electron beam which "paints" across the screen must return rapidly to its starting point at the upper left portion of the screen in order to begin the process again. This process occurs very rapidly (usually sixty times a second in VDTs made for use in the United States, and fifty times a second in VDTs made for use in Europe), and the energy emitted from the transformer by this process comes in short bursts, or pulses. As a result, the emissions of VLF and ELF are in the form of pulsed fields.

The effects of VLF (from 30 kHz to 500 Hz) and ELF (from 500 Hz to 5 Hz) electromagnetic radiation are not nearly as well understood as those of either ionizing radiation or other types of non-ionizing radiation. VLF and ELF radiation do not produce thermal effects except at high doses and prolonged exposures. As a result, these types of radiation were previously not thought to pose a health hazard at the levels present in common applications such as high power lines, hairdryers, electric blankets, and blenders.

Many radiation specialists claim that unless non-ionizing radiation produces a thermal effect (in other words, sufficient exposure to generate heat) it is not harmful. This is the criterion underlying most standards for exposure to non-ionizing radiation.[16] In the past few years, however, this thermal effect theory has been challenged by other scientists who claim that more subtle, but no less important, biological changes may take place after exposures to non-ionizing radiation below— sometimes far below—that required to produce thermal effects. These effects are said to be *athermal* or *non-thermal*. Some of the non-thermal effects which have been reported are behavioral changes (depression, anxiety), changes in the blood-brain barrier (which helps to protect the brain from toxins and viruses in the bloodstream), and physiological stress (see chap. 5 for a discussion of the stress response). The potential for changes in reproductive outcome has also been suggested. It should be remembered that biological effects may be temporary, and they may also be harmless or even beneficial. An important positive application of low-level non-ionizing radiation has already been recognized—the regeneration of bones in fractures which fail to heal normally.[17]

Non-thermal effects have been the subject of attention in the Soviet Union and other Eastern bloc countries for more than a decade, and were in fact the subject of some U.S. research in the early 1940s. Recent focus, however, has been the result of concerns about microwaves in consumer products use, telecommunications, and radar applications, and lower frequencies used in high power lines, satellite uplinks, and VDTs. According to W. Ross Adey, a leading researcher in bioelectromagnetics, "If nonequilibrium [non-thermal] phenomena are in fact important . . . cellular functioning can be significantly affected, and

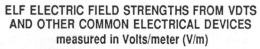

ELF ELECTRIC FIELD STRENGTHS FROM VDTS AND OTHER COMMON ELECTRICAL DEVICES
measured in Volts/meter (V/m)

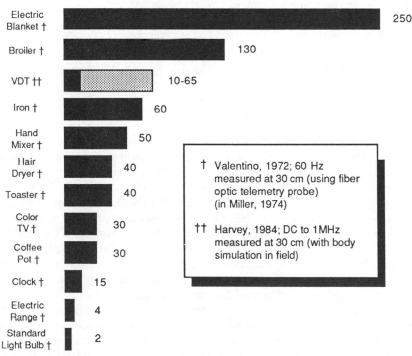

Device	V/m
Electric Blanket †	250
Broiler †	130
VDT ††	10-65
Iron †	60
Hand Mixer †	50
Hair Dryer †	40
Toaster †	40
Color TV †	30
Coffee Pot †	30
Clock †	15
Electric Range †	4
Standard Light Bulb †	2

† Valentino, 1972; 60 Hz measured at 30 cm (using fiber optic telemetry probe) (in Miller, 1974)

†† Harvey, 1984; DC to 1MHz measured at 30 cm (with body simulation in field)

Fig. 4. ELF Electric Field Strengths from VDTs and Other Common Electrical Devices

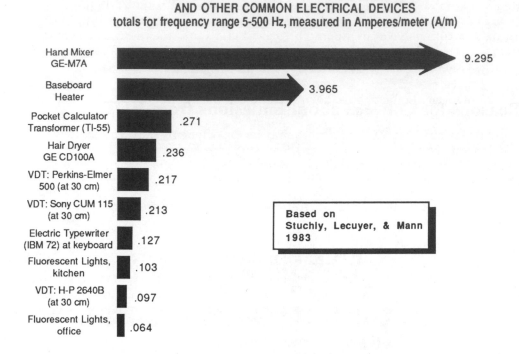

ELF MAGNETIC FIELD STRENGTHS FROM VDTS AND OTHER COMMON ELECTRICAL DEVICES
totals for frequency range 5-500 Hz, measured in Amperes/meter (A/m)

Hand Mixer GE-M7A	9.295
Baseboard Heater	3.965
Pocket Calculator Transformer (TI-55)	.271
Hair Dryer GE CD100A	.236
VDT: Perkins-Elmer 500 (at 30 cm)	.217
VDT: Sony CUM 115 (at 30 cm)	.213
Electric Typewriter (IBM 72) at keyboard	.127
Fluorescent Lights, kitchen	.103
VDT: H-P 2640B (at 30 cm)	.097
Fluorescent Lights, office	.064

Based on Stuchly, Lecuyer, & Mann 1983

Fig. 5. ELF Magnetic Field Strengths from VDTs and Other Common Electrical Devices

perhaps impaired, by athermal tissue interactions, which involve no appreciable change in temperature." The "if" in this hypothesis is a big if, and the controversy surrounding the recent developments in research into the non-thermal effects of electromagnetic radiation is, according to Adey, "one of the greatest conceptual revolutions in the history of science."[18]

Because of the lack of acknowledged thermal effects many governments have never established standard exposure levels for radiation in this frequency range. Guidelines for some portions of the VLF-ELF range have been adopted by professional associations such as the American Industrial Hygiene Association. The levels considered acceptable in the USSR and some Eastern European countries are considerably lower than those permitted in Western Europe and North America, although due to differences in measurement techniques and parameters these differences may be more apparent than actual.[19] In the range below 10 MHz, no guidelines have been established in the United States.

VDTs produce ELF radiation primarily in the 15-20 kHz range. Even though VDTs emit more radiation in this range than in any other except visible light, these fields are still at levels many orders below even the most restrictive Eastern European/Soviet standards, and are well below the level at which biological effects are currently known to occur. VDT emission levels are also far below the levels emitted by many other common devices used in the home, such as electric mixers and broilers (see figs. 4 and 5). This is, however, the range which has elicited the greatest research interest in connection with VDT health hazards, particularly as other recent research has added strength to the non-thermalist model. The pulsed nature of these emissions from VDTs has also been the subject of specific study, since research indicates that pulsed fields may be more capable of producing biological effects than non-pulsed (constant) fields at the same frequency. The most controversial and contradictory research findings have involved experiments investigating pulsed fields.

Reasons for Concern about Emissions from VDTs

Allegations that VDT operators may be experiencing certain health problems more often than other groups have led researchers to try to find a reasonable cause for the reported problems. Measures of emissions from VDTs have been of continuing interest because radiation emissions might theoretically provide such an explanation. Radiation emissions have been suspected in two particular health concerns, cataract formation and adverse pregnancy outcome.

Radiation Emissions and Cataract Formation

Cataracts are opacities in the lens which can cloud vision, sometimes to the point of blindness. They occur spontaneously in some people (particularly those over sixty), but the frequency of spontaneous occurrence in the general population is unknown. Cataracts which are severe enough to impair vision are often removed surgically.

Ionizing radiation is capable of producing cataracts. Exposures to ultraviolet, infrared, or microwave radiation at levels high enough to cause heating can also cause cataracts.[20] Cataracts caused by radiation have a very specific appearance which is unlike that of spontaneous cataracts. Called capsular cataracts, they form on either the posterior or anterior surface of the lens. Spontaneously occurring cataracts, on the other hand, appear in the lens itself. Thus, the position and appearance of a cataract indicate whether or not it could be radiation-induced. When radiation-induced cataracts are diagnosed, radiation exposures which might account for their development may then be explored.[21]

In 1977 two editors under the age of thirty-five, both of whom used VDTs at their *New York Times* jobs, developed what were diagnosed as radiation-induced cataracts. They claimed that radiation from the VDTs they used caused the cataracts. This incident began the Newspaper Guild's long involvement with VDT health and safety issues, specifically its concern that radiation from VDTs could cause cataracts and that VDT users might be at risk.

These two cases also led to the first survey of ionizing and non-ionizing radiation from VDTs conducted by the National Institute for Occupational Safety and Health (NIOSH). This survey found levels of all emissions to be below those known to have biological effects, leading some experts to reject out of hand the possibility that VDTs could lead to the development of cataracts.

Others, however, began to look at alternate forms of radiation emitted from VDTs. Concerns about long-term, low-level exposure to non-ionizing radiation, and about the presumed safe levels of radiofrequency radiation in particular, began to surface. However, the case for VDT-induced cataracts has steadily weakened, even though one worker in New York state was awarded workers' compensation for cataracts allegedly caused by radiation from her VDT. That decision has since been set aside, and a new trial will be held.[22]

One ophthalmologist, Dr. Milton Zaret, has claimed repeatedly that cataracts can be caused by radiation from VDTs.[23] The Panel on Impact of Video Viewing on Vision of Workers discounted his report of ten cases of VDT-induced cataracts on two grounds: 1) his reports on these cases had not been published in a refereed scientific journal, which suggests that his work may not have been measured against accepted scientific criteria; and 2) only four of the ten cases had significant visual impairment due to cataracts (many people have lens opacities which do not interfere with vision), and those four individuals were known to have other exposures or predispositions which were more likely to have caused their cataracts than VDT exposure.[24] The International Commission on Illumination also rejected Zaret's claims on similar grounds and concluded that "reports of cataract being caused by electromagnetic radiation emitted from VDUs should be considered as dubious."[25]

Zaret himself does not deny that some of his patients had had previous radiation exposures, but claims that the radiant energy cataracts he diagnosed in those patients were activated by the additive effects of exposure to non-ionizing radiation from VDTs long after the initial exposure to other radiation. He argues that current radiation standards are inadequate to prevent cataractogenesis (the formation of cataracts) in individuals susceptible because of previous exposures or certain illnesses.[26]

Other concerned vision professionals, such as optometrist Helen Feeley, point out that a number of factors can contribute to cataract formation, including lifetime radiation exposures, certain drugs, hereditary tendencies, and other ailments such as diabetes. Some drugs (such as tranquilizers containing phenothiazine) which produce light-sensitive reactions actually increase the body's absorption of ultraviolet light.[27]

However, other researchers claim that if the threshold for cataract-inducing exposure is very low or nonexistent, and if there are in fact "people who are more susceptible to radiation" than others, such individuals would be at greatest risk from general environmental radiation, which is considerably higher than the radiation exposure near a VDT. However, it is also true that occupational exposure standards do not claim to protect "100 percent of the workforce."[28]

In 1983, the Panel on Impact of Video Viewing on Vision of Workers released its report examining emissions from VDTs and reviewing research on VDT health effects of all types. They found no radiation hazards from VDT use. With respect to the question of cataracts, the panel reported: "We find no scientifically valid evidence that occupational use of VDTs is associated with increased risk of ocular diseases or abnormalities, including cataracts."[29] No other convincing data has surfaced to dispute this view, and at this time, the prevailing scientific opinion is that there is no link between VDTs and the formation of any type of cataract.

Radiation Emissions and Adverse Pregnancy Outcomes

When clusters of adverse pregnancy outcomes were reported in some VDT user populations, the search began for a cause, just as it had in the cataract investigations (see chap. 6 for a detailed discussion of reports of adverse pregnancy outcomes). It was logical to suspect radiation as a mechanism, because embryos are in a constant state of cell division, and some radiation can cause cell damage. Such damage may then be reproduced in new cells during division. Cell damage may have many effects on the human embryo or fetus, including spontaneous abortion, congenital defects, and predisposition to certain diseases such as leukemia. The effects usually are linked to the stage of development at the time of exposure. (Other effects, and other environmental factors and individual differences which have an impact on fetal development, are discussed in chap. 6.)

Ionizing radiation is known to be capable of causing such cell damage, and so ionizing radiation emissions from VDTs were the first to be suspected and investigated. But since levels of ionizing radiation were below even normal background radiation levels, investigators began to look elsewhere for possible explanations. The non-ionizing radiation emitted by VDTs thus became a cause for concern, and researchers began to study these forms of radiation, particularly the pulsed ELF and VLF electromagnetic fields, and to devise experiments with animals to investigate the possible effects of these emissions on reproduction.

Recent Research on the Effects of ELF and VLF Fields

Because emissions in this frequency range had not previously received as much research attention as ionizing radiation, less is known about safe levels of exposure to them. Further, the pulsed nature of the fields generated by VDTs was not accounted for in previous studies, and recent thinking indicates that biological effects may be dependent upon characteristics besides frequency, particularly the pulse of the field. These pulses are produced when energy from the flyback transformer is emitted in extremely short (millisecond) bursts. These bursts produce emissions which vary in intensity; while the frequency remains the same, the level (amount of emission) suddenly increases and then rapidly falls off, producing a peak-and-valley pattern. Different devices may produce pulses with different shapes; VDTs produce a pulse that is characterized by a sawtooth shape, rising very rapidly and falling off more slowly.[30] Standards have been based on constant (non-pulsed) fields. What is more, these extremely short pulses cannot be accurately measured by standard measuring instruments, which measure average (RMS) values, as if the energy level of the emission were flat, not variable. This makes the researcher's job more difficult. In addition, accurate measurement even of a continuous wave is extremely difficult in these low-frequency ranges because the wavelength is so long that the electric and magnetic fields must be measured separately. In experiments into the effects of these fields, the two components — electric field and magnetic field—are sometimes treated entirely separately. Thus some research refers to *PEMFs*, pulsed electromagnetic fields, and some refers to pulsed magnetic fields or *PMFs* only.

Researchers are now looking at these emissions to determine whether or not these special properties of ELF and VLF emissions have any significant effects on living matter, and if so, whether or not these emissions represent a significant risk to operators of VDTs or to their unborn children. Previously, it was assumed that the long waves of these low frequencies passed through human tissue almost without being absorbed. It was thought that there was no biological hazard unless exposures were very high, because only very high levels could lead to enough absorption to cause tissue heating. Since no harm was suspected, very little research was conducted on these frequencies and few national standards exist in this range.[31] Guidelines which have been promulgated by professional associations have been based on levels of exposure known to cause thermal (heating) effects.

Now these assumptions are being questioned and research on these emissions is currently very active. One study which generated a great deal of attention was performed by a Spanish research team (Delgado and Leal). Their 1983 report found that chicken embryos (fertilized eggs) which were exposed to PEMFs of a particular square wave shape showed a statistically significant increase in fetal malformations. This finding has come to be called the Delgado effect. This study, however, had two major drawbacks for those evaluating emissions from VDTs. The pulse shapes which were generated were not characteristic of the pulsed fields produced by VDTs, and the intensity and shape of the magnetic pulses were not well controlled, resulting in the exposure of the embryos to some types of waveforms which were not part of the original research design. In spite of this, the study was found to be sound by Rick Tell of the Environmental Protection Agency, who went to Spain to investigate Delgado's laboratory.[32]

Delgado's findings have led other researchers to attempt to replicate these results, which are certainly of interest whether or not they are directly applicable to questions involving VDTs. In 1984 Dr. Kjell Hansson Mild, of the National Board of Occupational Safety and Health, Sweden, reported that he had replicated the Delgado experiments and confirmed the general effect, "but with a lower rate of malformations."[33] Project Henhouse, a multi-laboratory project attempting to replicate the Delgado effect, is currently being conducted jointly by the EPA and the Office of Naval Research. Testing is expected to take a year.[34]

In at least one study on mice (Tribukait et al., Karolinska Institute, Stockholm 1986) rectangular fields similar to those used in the Delgado study did not produce any significant adverse effects on the fetuses, but a sawtooth-shaped pulse which closely resembles the fields produced by VDTs did result in a statistically significant increase in fetal malformations. More importantly, there were similarities among the types of malformations. Such patterns of similarity often indicate a common cause for the malformations. These researchers concluded that "both the frequency and the type of malformation found after exposure to PMF with sawtooth pulses during early pregnancy including the whole period of organogenesis [organ development] might indicate that PMF of this specific type has teratogenic [birth defect causing] effects."[35]

The preliminary results of this experiment received a great deal of attention when they were released just prior to the Work with Display Units conference in Stockholm in the spring of 1986. The experiment was strongly criticized because the size of the sample was small and the control group (unexposed mice) had fewer malformations than was expected for that particular strain of mice. The investigators have since increased the number of subjects by around 75 percent. This expanded study yielded a malformation rate in the control group that was closer to the expected rate for that strain, and still found a statistically significant increase in external malformations in mouse fetuses exposed to one of the sawtooth pulses. A higher (although not statistically significant) rate of malformations in mouse fetuses exposed to another of the sawtooth pulses was also found.

No differences were observed in the rate of resorptions (aborted embryos which were assimilated by the mother's system) and fetal deaths, nor in the body weights of fetuses in the two groups. These rather unexpected findings caused one of the researchers to regard the malformation findings with caution. Other observers also point out that the statistical analysis which yielded significant results was performed only on malformations, and not on the results of the combined effects (body weights, resorptions, and spontaneous abortions, as well as malformations). Usually individual effects like malformations are calculated only if the overall results are in themselves statistically significant. However, researcher Bernard Tribukait from the Swedish team defends their analysis as appropriate for the study as designed. This study has been replicated in Uppsala, Sweden, at the University of Agriculture. Only very early news reports were available for review, and it is not clear from these reports how comparable the Uppsala and Karolinska studies are. A recent article in *VDT News* quotes Uppsala researcher Walinder: "We are seeing a very early effect that makes animals more susceptible to implantation. Embryos that are normally rejected by the mice were accepted." [36] Until the full report is available for examination (it is presently available only in Swedish) it would be premature to comment on findings or experimental differences. Both the Karolinska team and the Uppsala team are planning new experiments. [37]

In a different experiment, the results of which were also presented at Stockholm, Polish researchers exposed male and female laboratory rats to radiation from black-and-white TVs and from color TVs. They claimed that the female subjects had a higher number of involuted fetuses (resorptions) and lower litter weights (particularly those exposed to black-and-white TV), but they did not state the level of statistical significance. They also found a reduction in testicle weight in the male subjects. [38]

These same researchers also examined groups of female VDT and non-VDT operators at two locations. They found that at one location, where the workers were performing tasks that involved a heavy mental load, the VDT operators suffered from "more frequent spontaneous miscarriages" as well as more menstrual problems, more headaches, and more generalized symptoms than the non-VDT operators. At the other site, where the task involved "simple operation on numerical data," the VDT operators showed no more problems than the two groups of clerical non-VDT users. Again, the level of statistical significance was not reported. [39]

The results of these two Polish studies also received much media attention in reports from the Work with Display Units conference. However, methodological problems in both experiments—including failure to correct for known confounding factors and poor experimental control in general—make the conclusions from these studies highly suspect, and other researchers have generally rejected these studies as scientifically unsound.

In Finland, a chicken egg study conducted by Jukka Juutilainen found weak PEMFs of three different types to be biologically active in the frequency range above 30 Hz (in the ELF portion of the spectrum). Still other studies on chicken embryos (Sandstrom and Mild in Umea, Sweden, also reported at Stockholm) have shown no statistically significant differences in malformations, even though the pulses closely matched those used in the mice study at the Karolinska Institute. However, Sandstrom and Mild themselves pointed out the difficulties in comparing their results to the experiments of Juutilainen and

SAINT PETER'S COLLEGE LIBRARY
JERSEY CITY, NEW JERSEY 07306

others because of differences in experimental procedures, including the orientation of the eggs, which can change the amount of radiation actually absorbed by the part of the egg containing the fetus. According to *Microwave News*, "A consistent picture of the effects of weak pulsed electromagnetic fields (PEMFs) on developing chick embryos continues to elude researchers."[40]

Current and Future Directions in Research and Standards

The recent experimental findings from Sweden, along with the continuing attempts to replicate the Delgado effect, have kept the radiation controversy alive, and have called into question the current standards for non-ionizing radiation exposure and how they are determined. In fact, for the frequencies under examination (below 300 kHz), there are no government standards, and there are not even any recognized guidelines for frequencies below 10 kHz.

The increased interest in these extremely weak, low energy fields has also brought to light much of the work on ELF and VLF radiation done in the USSR and in Soviet bloc countries, work which is based on different research assumptions than U.S. research. Soviet work has for some time attributed non-thermal effects to non-ionizing radiation, including effects on the central nervous, cardiovascular, and endocrine systems.[41]

Radiation research in the United States has traditionally been directed toward the discovery of gross effects (pathological damage such as birth defects), and standards have been based on complicated formulas of risk assessment.[42] It is important to recognize that while standards draw upon the results of scientific research, they do not rest entirely upon those results. They are, in fact, policy decisions which weigh costs — to industry, to government, to society, and to the individual — against benefits. The purpose of occupational standards for exposures to hazardous agents is to balance the risk of effect against the benefit to society, based on our understanding of the following:

1. What level of safety is "safe enough"?
2. How much is society willing to pay for safety?
3. How can the "acceptable risks" be equitably distributed?
4. How reliable are our methods of risk measurement?[43]

Traditional risk assessment theories used in the United States (and in many other countries) have held that occupational exposure could be higher than public exposures, because occupational exposures occur to a voluntary population of healthy adults who are informed of the risk, who are compensated for these exposures in higher wages, and who are more likely to be monitored for health effects.[44] These standards may be altered as our knowledge of the risks/benefit ratio increases. In fact, in 1985, the American National Standards Institute (ANSI) standard C95.1 for microwaves lowered the maximum exposure limit in apparent accord with concepts current in the Soviet Union for a decade or more.[45] According to William Murray of the radiation section at NIOSH: "Basically the question is whether or not the present standards are adequate. The standards for occupational exposure are never as good as they could be. You never have enough information to set a perfect standard."[46]

While microwaves are not at issue in the VDT debate, the revision of microwave standards may be an indication that a new era of research in non-ionizing radiation is just beginning, and that as a result new standards reflecting the views of the non-thermalists may be in the offing. According to W. Ross Adey, "Safety standards based on nonequilibrium [non-thermal] considerations would require the acceptance of sharply lower exposure levels than would thermal standards."[47]

In the meantime, it is not clear how VDT manufacturers will proceed, or how unions, governments, and users will respond during the interim period of further research. The World Health Organization (WHO), in its 1985 working document, stated that VDT emissions cannot be regarded as hazardous and that the available data do not support claims of links between VDT use and reproductive harm. At the same time, WHO acknowledged that possible radiation effects cannot be completely ruled out.

Sweden now insists that VDT manufacturers eliminate electrostatic (static electricity) fields and substantially reduce low-frequency electromagnetic fields in order to sell their terminals in Sweden. The Swedish civil service and numerous local and private entities have made agreements which provide for the transfer of pregnant women to non-VDT work without loss of pay. Similar agreements have met with varying degrees of acceptance in the United States, Canada, and elsewhere. Swedish Labor Minister Anna Greta Leijon, in an opening address at the Work with Display Units conference, said: "Research and development . . . is a very time-consuming process, and it will probably take us a very long time to find the answers to all our questions concerning possible health hazards and causal connections. Meanwhile we cannot sit idly waiting for the research findings to come through. If we really mean what we say about taking people's apprehensions seriously, we must also be prepared to take unconfirmed but suspected risks into account. . . . Although there is no evidence today of this radiation having any biological effects, there is no reason for burdening the occupational environment with factors which can be eliminated."[48]

Shielding to reduce or eliminate ELF and VLF emissions from VDTs has been suggested as a viable approach elsewhere, as well. In fact, the electric component of the fields in these frequencies can be fairly cheaply shielded, but the magnetic component is harder to reduce.[49] According to a report for the Office of Technology Assessment by Robert Arndt and Larry Chapman, "Where reductions in exposures are reasonably attainable, experience with other agents has proven the value of this approach, especially when uncertainties exist." They also point out that should these low-level radiations be found to have biological effects, the impact will extend far beyond the VDT alone: "They extend to fields emitted by other office equipment (lights, copiers, heaters, facility wiring, etc.) and consumer products, as well as broadcast communications and power transmission technologies."[50]

Understanding Radiation Research: A Layperson's Guide

Research into the effects of radiation is ongoing and results remain inconclusive. Media reports are sometimes biased, with the same study results presented differently in different publications. How can the intelligent, concerned non-scientist make an individual judgment about the importance of study results? One method is to weigh differing reports according to the reliability of the source. A more comprehensive method is to read the original study results or the summary report. With a modest amount of preparation it is possible for the layperson to make sense of such reports and to make informed judgments about their value. Lack of technical expertise should not prevent a layperson from adopting a critical stance toward the often contradictory results of research.

Research of the type used to investigate the direct effects of radiation is generally composed of laboratory experiments of a particular type. The mechanism or cause, radiation, is proposed, and the investigators attempt to determine the effects of specific exposures on subjects. They use a controlled technique for exposing some non-human subject (usually pregnant mice or fertilized chicken eggs) to the suspected cause (i.e., the specific type of radiation). Researchers then observe what happens as a result of the exposure and report the relative importance of the results in terms of *statistical significance.*

Statistical significance. Because experimental subjects, whether human or animal, are different from one another in many ways, it is important for researchers to be able to tell which differences are due to normal or random variations and which are due to an experimental procedure. Statistical tests are the tools that scientists use to make these distinctions. Most (but not all) commonly used statistical tests generate a number, called *p* or *p value* or *probability*, which expresses the likelihood that the observed differences between experimental and control groups are due to normal or random variation. For example, a $p = .3$ value means that 30 percent of the time, the results could be due to chance and not to the experiment. Researchers customarily do not accept p values above .05 as statistically significant, and probabilities below .05 are considered even stronger indications of a link between the experimental condition and the results. Probabilities may be reported in any of several ways:

$p = .01 \rightarrow$ observed results could occur by chance 1 percent of the time

$p < .01 \rightarrow$ observed results could occur by chance less than ($<$) 1 percent of the time

significant at the .01 level $\rightarrow$ observed results could occur by chance 1 percent of the time.

A statistic is merely a number, and it is only as good as the work that produced it. The critical reader will look beyond a claim of statistically significant results at the whole scheme of the research, examining such features as control of variables and appropriate selection of control groups.

In evaluating results of animal studies, even those which are highly significant, it is crucial to remember that applying such results to humans is risky. While responsible researchers use animal species which are appropriate (for example, cataract experiments frequently use a particular strain of rabbit, because that particular rabbit's eyes are more like human eyes than any other easily managed laboratory animal), there are inescapable differences between species. Usually, when one animal study is highly significant, other researchers perform the same experiments and try to reproduce (replicate) the same results, first with the same species and then with others. The more times that the results of the experiment are the same in different appropriate species, the more likely it is that those results may be generalized to humans.

The layperson, then, can use the statistics which are reported as a way to determine the apparent degree of importance of the findings. Beyond the statistics, the reader can look at the experimental design and methods: How large were the control and experimental groups? How appropriate were they? How well-controlled were the variables? Were the variables appropriate to the question the reader is interested in? For example, some radiation study findings which have been discussed in the context of VDTs did not, in fact, investigate radiation like that emitted from VDTs.

At first this might seem to be a difficult task for the non-scientist, but it is in fact a skill that may be developed by reading critiques of study findings and by gradually becoming familiar with the design, methods, and reporting techniques of such experiments. It is then possible to make informed judgments based on first-hand knowledge, rather than on other interpretations which may be biased in either direction.

Conclusions

While the controversy regarding allegations of adverse health effects of radiation from VDTs is not yet resolved, there is now general agreement on some points.

Repeated measurements of VDTs have never shown any amount of ionizing radiation which could be considered harmful to humans or human fetuses. Lead shielding of VDT operators is thus unnecessary and ill-advised. Neither is there any evidence that cataracts are caused by radiation from VDTs.

Non-ionizing radiation is emitted from VDTs, but all levels are greatly below any levels which are known to cause harm to humans or human fetuses.

Reports of adverse pregnancies, however, have led to careful examination of ELF and VLF radiation, and a great deal of new research has been and is being performed to investigate the effects of these frequencies. These frequencies were previously thought to be harmless except at very high levels, and little research had been done.

So far, the results of research on ELF and VLF fields have been mixed, but even the experiments which have claimed significant results must be regarded carefully, since they used mice or chicken embryos as subjects, and we cannot judge for certain how well these results translate to the human organism, and also since the fields in question may not be comparable to those from VDTs. If further research shows that these frequencies can have adverse effects on human cells at weak exposures (below levels which heat tissue), the impact of such findings may be substantial and may lead to significant revisions in exposure standards.

At this time, however, most responsible researchers seem to indicate that adverse effects of these fields, if they are found to exist, are likely to be very small in the case of VDTs, since they emit less radiation in this frequency range than many other common devices, such as hairdryers and electric blankets. VDTs and other common appliances which emit ELF and VLF fields are likely to be found safe even if exposure standards are lowered. More caution would be warranted in exposures to sources of higher intensity ELF and VLF radiation such as satellite uplink stations and high power transmitters.

Nevertheless, until research on these fields establishes a clearer pattern, some operators may experience real anxiety about these issues. Such anxiety may be stressful, and these operators' concerns should be treated seriously.

Further Readings

Adey, W. Ross. "The Energy around Us," *The Sciences* 26 (1) (Jan/Feb 1986), 53-58.

Slesin, Louis. "People Are Antennas, Too: The Biology of the Electromagnetic Spectrum," *Whole Earth Review* (50), 50-55.
While these two articles are not specifically about VDTs, they provide good introductions to the biological effects of non-ionizing radiation for lay readers. Adey summarizes the thermalist vs. non-thermalist debate. Some background in physics or biology is helpful in understanding his somewhat technical presentation. Slesin provides a more popularized presentation of the same debate.

Arndt, Robert and Chapman, Larry. *Potential Office Hazards and Controls.* Contractor's Report Prepared for the Office of Technology Assessment. University of Wisconsin. 1984.
The full text (160 pages, plus references) of the authors' report to the OTA provides a very thorough introduction to the issues of office health concerns in general and VDT health concerns in particular. The discussion of radiation emissions is especially straightforward and accessible.

The Koffler Group. "Radiofrequency Radiation and Video Display Terminals." *Office Systems Ergonomics Report* 3 (5) June 1984.
This special issue summarizes the positions of most of the active participants in the debate and presents a brief discussion of the scientific controversy, including some comments on non-thermal effects. The results from the Delgado team are discussed and evaluated. A chronology of events is included.

National Research Council, Committee on Vision. *Video Displays, Work and Vision.* National Academy Press. 1983.
Numerous studies have reported measurements of emissions from VDTs. Rather than citing these studies, the reader is referred to this report, which is the result of a National Research Council review requested by NIOSH. The most frequently cited emissions studies are reviewed in Chapter 3. This chapter also discusses the biological effects of radiation, and includes a lengthy discussion of the cataract issue. General conclusions about radiation hazards from VDTs are presented. This comprehensive report also discusses visual, musculoskeletal, and stress complaints of operators, summarizes relevant research, and presents recommendations.

3   Eyes and Vision

- Eyestrain is the single largest category of health complaint among VDT users.

- There is no evidence to indicate that VDT work leads to any permanent visual damage.

- Introducing VDTs requires some changes in workstation design and workplace lighting in order to control glare and reduce visual complaints.

- Following design guidelines in the selection of monitors is also important in reducing visual complaints.

- Regular eye examinations are recommended. About one-third of the workforce may have uncorrected or undercorrected vision problems.

- Aging operators may experience particular visual difficulties. Glasses used to correct age-related visual changes may need to be specially made to bring the display into focus at normal VDT viewing distances.

- Contact lens wearers may find that the typically drier VDT environment is uncomfortable, and may experience further drying due to reduced blink rates.

Eyestrain is the single largest category of health complaint among VDT users. Studies of visual discomfort among VDT operators regularly report that more than half of the operators experience eyestrain or related visual discomfort,[1] and some studies report that as many 94 percent of the operators suffer from visual complaints.[2] In fact, these complaints among VDT operators are so common that some have suggested that VDT really stands for "Visual Discomfort Terminal."[3]

There is no question that VDT work is visually demanding, and that some existing visual problems may be exacerbated by VDT work. However, many researchers claim that VDT work is not intrinsically more visually demanding than many other occupations which require prolonged close attention to a visual task, and that therefore there are no new visual problems inherent to the VDT task.[4]

According to Bell Laboratories researcher Steven Starr, "The VDT does not seem to be a major new source of discomfort in the workplace. It does not stress the visual system more than analogous near-vision work done without VDTs, and its use need not reduce job satisfaction."[5] Nonetheless, VDT operators do experience eyestrain with great frequency. According to the Panel on Impact of Video Viewing, in most surveys over 50 percent of VDT operators report some visual discomfort, and these complaints are generally more prevalent among VDT operators than non-VDT workers.[6] But visual complaints are common among clerical workers in general, and other researchers have found that VDT operators do not report eyestrain any more frequently than other clerical workers performing similar close visual work.[7] Still others, however, have reported significantly more complaints from VDT users compared to other clerical workers, particularly among those using VDTs for more than four hours per day. Intensive data entry seems to generate more complaints than other VDT tasks.[8]

Visual complaints are seldom the result of one condition. According to some researchers they are a combination of individual differences in visual acuity (the ability to discriminate fine detail) or eyeglass correction and poor visibility conditions in the display (flickering, dazzling, and reflection) and in the objects which the operator views (poor hardcopy and display legibility).[9] In addition, the visual system may be the first to display symptoms of discomfort or fatigue, even though the source of discomfort may be postural.[10]

VDT Work and Visual Complaints

Eyestrain (also referred to by its technical name, *asthenopia*) is not a visual defect but is rather a catchall description for various symptoms of visual discomfort, including burning, itching, tiredness, aching, soreness, and watering. Other diverse symptoms of visual disturbance such as blurry vision or altered color perception have also been reported in connection with VDT work. Some operators find that their eyeglass prescription must be changed after some period of work with VDTs, or that they need glasses for the first time. Contact lens wearers may complain of itchiness, dryness, and discomfort when they wear their contacts while working on VDTs. All of these symptoms have been reported by VDT operators, and in many cases the VDT has been blamed. By far the most serious allegation is the persistent claim of a causal relationship between radiation from VDTs and the development of cataracts. Prevailing opinion is that no such connection exists (for a thorough discussion of this question, see chap. 2).

One reported symptom of VDT viewing, an altered color perception or after-image, may be particularly disturbing to operators unless they are aware of its harmless nature. This after-image is actually a well-known phenomenon which may occur when a viewer looks at a colored object steadily for a period of time and then shifts the vision to a light backdrop. The backdrop will appear to take on the color opposite (or complementary) to the steadily viewed object. Thus, continued viewing of a green monitor may lead to the perception that things look pink. This effect disappears after a short period of time and does not harm vision in any way.

The aging process naturally brings on certain visual problems such as presbyopia, which is a reduction in the ability to focus on near objects caused by decreased flexibility of the lens (commonly called farsightedness). This may pose special problems for the older VDT operator.[11] Presbyopia is usually corrected by bifocals, trifocals, or reading glasses, and the correction is established to bring near vision into sharpest focus at the normal reading distance of about twelve inches. Since typical VDT viewing distance is generally in the range of twenty inches, presbyopic operators may require special eyeglasses that are corrected for the increased distance. They may also require a larger, higher near lens than is typically prescribed for bifocal wearers, so that the head does not have to be tilted back at an unnatural angle in order to see through the near lens. There are other solutions as well. Some bifocal wearers may choose to view the screen through the upper, far vision lens, depending on viewing distance, while others may prefer to use single correction near vision glasses for VDT work.[12]

Contact lenses may also pose special problems in the VDT environment. Dry environments and the attendant static attraction of dust to the screen area may make contacts uncomfortable, and the heat generated from VDTs tends to make this environment drier than the equivalent non-VDT office unless steps are taken to increase the humidity.[13] In addition, many VDT operators show a marked tendency to decrease their blink rate. The drying effect of reduced blinking may be particularly troublesome for contact wearers. Conscious blinking effort, perhaps supplemented by eye drops which simulate normal saline (but not medicated drops, which can have adverse effects when used too frequently), may reduce this problem.[14] Contact wearers are also subject to all the other visual concerns discussed here, including the itching, burning, and soreness of eyestrain, and some operators may find that they prefer to wear eyeglasses at work instead of contacts. Cleaning contact lenses frequently with protein cleaners may help prevent dry spots caused by protein deposits, though some wearers may be sensitive to such cleaners.[15]

There is no indication that VDT work leads to any permanent damage to the visual system; the symptoms which have been reported are all of a short-term nature. It also seems clear that very few of these complaints are directly related to VDT equipment per se, but rather that a number of them are connected to the type of work performed, the work environment, and how VDTs have been introduced into the workplace.

Assessing the Effect of VDT Work on Visual Comfort

While there is general agreement that VDT operators report a number of valid visual complaints, there is at this time no generally agreed upon way to measure eyestrain or visual fatigue. There are very sophisticated techniques for measuring specific functions of the visual system, such as the number of eye movements made while reading copy on paper and copy on screen display for the same task; for such functions as accommodation and convergence; and for specific visual defects such as presbyopia, myopia, and astigmatism. However, researchers do not agree about what to measure when the problem under investigation is eyestrain. Some researchers claim that measuring certain visual functions such as accommodation, visual acuity, convergence, and muscle balance can be used to determine the extent of visual fatigue,[16] while others counter that there is no apparent connection between the frequency of subjective complaints—the reporting of symptoms—and actual visual screening results.[17]

Until an accepted measure of eyestrain is available, researchers must rely on subjective reports from operators for their data.[18] Such reports often take the form of surveys or questionnaires. Comparing the results from different surveys can be difficult for two reasons. First, if the surveys were performed in environments which differ greatly, the data cannot be compared reliably. Second, the interpretation of surveys varies widely; a survey which accepts a positive response to a statement such as "I feel headaches sometimes" will present very different results from one which only computes positive responses to a statement such as "I always experience headaches." This difference in interpretation could alter the incidence of reported headaches in the same population by as much as 40 percent.[19]

Questionnaires are also notoriously subject to reporting bias, which causes subjects to be more likely to fill out and return questionnaires if they have some reason to think that the results of the study are personally relevant. For example, a questionnaire about eyestrain may get a lopsided number of responses from subjects who experience eyestrain. Underreporting and collaborating with other subjects are also forms of reporting bias. Thus it is important that questionnaires and surveys be carefully designed and administered by experienced investigators. In reality, however, they are the most misused of all research instruments. How many offices have invested in chairs after conducting an informal survey which asks, "What do people think of these chairs?"

In research on the effects of VDT work, it is important to match the type of task, amount of use, and environment for the group using the VDT as closely as possible to the type of task, amount of use, and environment for the group not using the VDT. An experiment which compares operators in a word processing pool with private stenographers taking dictation would not be very well matched. This has proven to be the hardest part of choosing control groups, because VDT work is often organized differently than clerical work. The researcher must also control for other differences which might have an impact on the outcome of the experiment, such as age, visual correction, and other individual differences between members of each group. These are potentially confounding factors which make it difficult to evaluate the reliability of experimental findings.

Certain special problems are inherent in field studies. Field studies cannot be as carefully controlled as laboratory experiments and, therefore, the results obtained from them must be evaluated with caution. The difficulties involved in administering and interpreting surveys and questionnaires have already been discussed. But there are some questions which simply cannot be examined effectively in a laboratory setting, and others in which the real world applicability of a field study is clearly preferable to the manipulated environment of a laboratory experiment. Much research on the effects of VDT use is of this type, and the results of such studies must be evaluated carefully by weighing the appropriateness and reliability of each study's methods and design.

Reducing Visual Discomfort in VDT Work

In spite of the fact that measurements of visual fatigue are not as precise as researchers would wish, it is clear that eyestrain is the most frequent complaint among VDT workers. It is also true that even though the extent of the problem is not known there are a number of actions which can be taken to improve visual conditions for VDT work.

In addition to the specialized needs of bifocal and contact wearers, visual discomfort among VDT users in general may be significantly reduced by giving close attention to the design of the monitor and display, ambient and task lighting, any resultant glare problems, and operator vision correction. Rest breaks, proper workload, and good arrangement of working materials are also important in reducing visual fatigue.

Monitor and Display Design

Some aspects of visual discomfort and fatigue can be alleviated by improved design and installation of the monitor itself. The actual design of the monitor and its display can help reduce both visual and muscular fatigue, and thus several design features need to be considered when selecting a monitor. These are the adjustability of the monitor itself, the quality of the display, and the polarity of the display.

Monitor Adjustability. It is essential for the monitor to have tilt and swivel adjustments.[20] This is important not only to permit more comfortable head and neck positions, but also because adjustments of monitor position can be crucial in controlling glare. Glare has long been considered a factor in eyestrain. The user may also change posture to try to avoid glare, which can lead to postural discomfort. "The best measure against specular reflections [glare] is . . . the adjustability of the screen (tilt and turn) which can make most reflection invisible to the user."[21] This adjustability need not, however, be built into the monitor itself; an add-on pedestal support can provide a good solution, as long as it does not make the screen too high.

Display Quality. A significant body of work exists regarding the technical aspects of display design. Many guidelines are more applicable at the design stage than at the point of selection and use, and they are not discussed here. Technical discussions about such specifics as modulation, use of color, linearity, font design, and measurements of luminance, contrast, legibility, readability, and strokewidth may be found in the titles selected for further reading. The following points are discussed as they apply to the selection and use of display terminals.

a. *There should be no apparent flicker.*

When the phosphors on the inside surface of the CRT are excited by an electron beam they glow, producing the display. The beam scans across and down the screen at a fixed rate (usually 60 cycles per second, or 60 Hz). The glow of the phosphor remains briefly after the beam has moved on—the time span of this glow is the phosphor's *persistence* rate. The rate at which the beam repaints the screen is the *refresh* rate. If the refresh rate is too slow or the persistence time is too short, the characters may fade away before being refreshed. If this occurs it is perceived as *flicker.* The point at which flicker is apparent, called critical fusion frequency (CFF), varies markedly from person to person, but standard U.S. refresh rates of 60 Hz seem sufficient to prevent the perception of flicker for most viewers when the display polarity is lighter character/darker background (see discussion below). Darker character/lighter background displays (reverse video) require a refresh rate of at least 80 Hz (preferably 90 Hz) in order to overcome apparent flicker, since a larger area of the screen is actually luminous. Flicker is more apparent when it is seen with the peripheral vision rather than in the direct line of sight.

b. *Brightness and contrast must be adjustable.*

These controls allow the user to make adjustments to optimize the legibility of a display in different lighting environments and also to obtain the sharpest display given individual differences in acuity, viewing distance, and so on. Both the ANSI draft standard and the guidelines of the International Trade Union Federation stipulate that brightness must be adjustable. Contrast adjustability is also desirable (ANSI standard) or required (trade union guidelines).

c. *A minimum 7×9 matrix should be used for text processing applications.*

The characters on a display are not solid lines, but rather a series of dots (called pixels) which the eye blurs together to produce the perception of a solid line, rather like a connect-the-dot picture. The number of dots available to produce a character within a single character space will determine how readily the eye makes this merger. The more dots, and the less space between dots, the more solid the characters appear. The pattern of dots available to produce a character is called a *matrix*; a 7 × 9 matrix would be seven dots wide and nine dots high (see fig. 6).

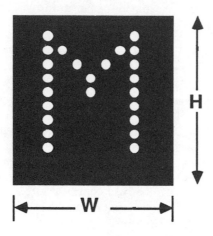

Fig. 6. Dot Matrix of Letter M

d. *Character height, width, design, and spacing should enhance legibility and readability.*

Measurements for legibility (the identification of single characters) and readability (the identification of words and word groups) are presented in most standards. For more information see section 6 of the ANSI draft standard, *American National Standard for Human Factors Engineering of Visual Display Terminal Workstations.*

e. *The display should have high resolution.*

Resolution is a measure that describes the sharpness of the display. The characters should appear crisp, not fuzzy. The brightness and contrast of the screen should be adjusted to achieve the sharpest display possible before evaluating screen resolution. Metrics used for evaluating resolution are presented in section 6 of the ANSI draft standard.

Display Polarity. The most common type of VDT display uses lighter

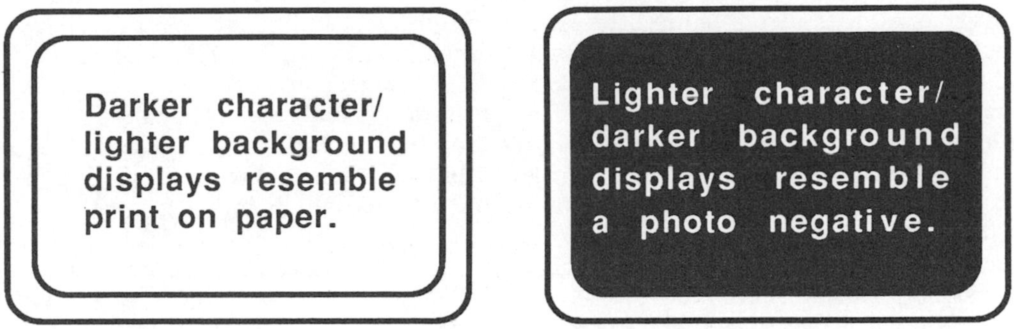

Darker character/ lighter background displays resemble print on paper.

Lighter character/ darker background displays resemble a photo negative.

Fig. 7. Two Different Display Polarities

characters on a darker background. Many newer VDTs allow the operator to choose between this lighter character/darker background (standard) display and a display which more closely resembles the printed page, that is, darker characters displayed on a lighter background, also called reverse video (see fig. 7).[22]

There is a continuing debate about which display polarity is preferable. It is fairly well agreed, however, that higher ambient light levels may be comfortably used with reverse video, because specular reflections (glare) are less apparent than when a lighter character/darker background display is used.[23]

Ideally, the system should allow the operator to choose either polarity, and such displays are becoming more common. Fast refresh rates were not readily producible in the first CRTs, so lighter character/darker background displays were originally adopted as the norm because they did not produce apparent flicker at normal refresh rates of 60 Hz (U.S.) or 50 Hz (Europe). Refresh rates of at least 80 Hz are necessary to prevent perceived flicker in reverse video displays, because such a large portion of the display must emit light. Today displays with refresh rates of 80 Hz or higher are commercially available, although they are more expensive than the earlier technologies. Other refresh techniques are becoming common, among them interlacing, which can double the apparent refresh rate by refreshing alternate lines.

Lighting

Several studies, including a Swedish study by Hultgren and Knave and several NIOSH studies, indicate "a recurring pattern of discomfort from lighting problems in VDT workplaces."[24] Similar lighting-related complaints are also common among typists, "although the causes may be different."[25]

The introduction of the VDT frequently creates glare problems that are not present in the paper-based office. Unlike most of the materials in a traditional paper-based office, the terminal has a reflective glass surface. Bright white paper reflects light (hence the green, blue, or yellow tints in many steno pads), but not to the extent that a display screen does. The screen's vertical orientation means that it will pick up reflections and direct light from above the operator (overhead lights), from behind the operator (light streaming through a window), even from an operator's light-colored clothing (reflected light). The resulting glare can be as irritating and tiring as driving into the sun without sunglasses. When glare is annoying and adds to the operator's visual load, it is called discomfort glare. The result can be distraction, headache, and visual and physical fatigue. When glare is severe enough that making out the information on the screen becomes difficult or impossible, it is called disability glare.[26] Although the precise role which glare plays in eyestrain is not well understood, both types of glare can reduce productivity and increase error rates, as well as cause operator discomfort.

Glare Control. While glare is one of the most widespread problems in VDT environments, it is also one of the most correctable. There are two approaches. One is to remove the cause of the glare. The other is to apply an antiglare measure to the screen. By far the most effective, and often the least expensive, is to remove the cause of the glare (see fig. 8). This can be done in a number of ways, depending on the cause. If sunlight striking the screen from behind or striking the operator from in front is the prime cause, the VDT can be moved parallel to windows. If that is not possible, or if that creates other glare problems, drapes or other window treatments can be used.

Similarly, moving the VDT so that it is parallel to and not directly under overhead lights can eliminate that source of direct glare. Baffles and louvers may also be used to break up direct overhead lighting. Pedestals for the monitor which allow it to tilt backwards and forwards and/or swivel side to side can help reduce both direct and reflected glare, though care should be taken that the use of a pedestal does not raise the monitor height to an uncomfortable level. Other approaches can include using indirect lighting or reducing overhead lighting supplemented by task lighting.

Other sources of glare are reflective surfaces in the office, including walls, desks, and keyboards. Potentially reflective surfaces in the VDT work area should thus have matte finishes.

If glare cannot be eliminated at the source, then an antiglare measure of some type may be appropriate. Various kinds of glare filters are available and they have received a great deal of attention. These filters can reduce some problems created by glare, but not without introducing other factors which must be considered. Some filters degrade the quality of the display noticeably by reducing the luminance as much as 65 percent.[27] Other filters are especially prone to collecting dirt, which further obscures the display. Some are quite effective but disproportionately expensive. Many filters now advertise that they reduce static electricity by the addition of a ground wire, and the reduction of static may help to reduce the attraction of dust particles to the filters. Antistatic mats with grounding wires either for floors or as keyboard mats can be used to solve the static problem without the display degradation problems associated with glare filters. Because of the problems associated with filters they are recommended only if other methods of controlling glare are not practical. The same is true of hoods, which can be homemade or commercially produced. They may block a glare source effectively, but they often require the operator to maintain a fairly constrained posture which can create fatigue and discomfort, and they are not effective against glare resulting from reflection directly in

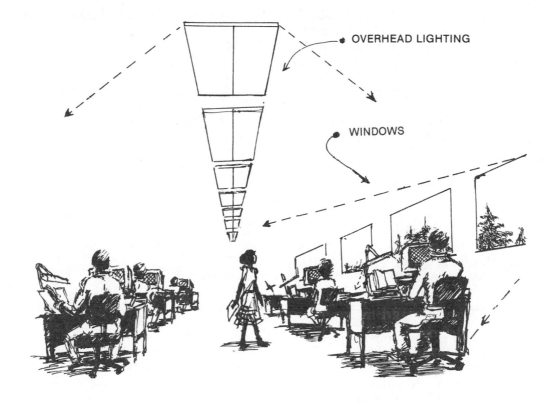

OVERHEAD LIGHTING

WINDOWS

Fig. 8. Reducing Glare at the Source

front of the screen, including reflection from the operator.[28] Many
manufacturers now use etching or coating methods to reduce monitor
reflectance.

Office lighting itself can contribute to poor VDT viewing conditions. The
paper-based office task has generally led to a more-is-better approach to office
lighting, but for the VDT-based office this is not the case. Overall luminance in
the typical office has been at least 500 lux (and sometimes much higher), but for
VDT-based offices recommended luminances are much lower, around 300 lux
(for discussions of lux and other measurements of lighting, consult the sources
listed in "Further Readings"). These lower light levels often help reduce glare
but may make it necessary to provide task lighting to illuminate hardcopy
source documents.

In repeated studies office workers show a strong preference for natural
light as an illumination source. Unfortunately, it is not practical to use only
natural light, both for seasonal reasons and because of the restrictions it would
place on architectural design. In fact, even the areas of the office which can
receive a substantial amount of sunlight are not perfectly suited for office work.
Shades and other window treatments are often necessary to control glare,

particularly where VDTs are installed. Solar heat passing through large window areas, especially those that cannot be opened, can add substantially to the thermal load, and in the winter heat loss through glass windows can be great. As a result most offices must rely heavily on artificial lighting, often in combination with some natural light.[29]

Fluorescent Lighting. Controlling glare and optimizing office lighting are intrinsic to improving the quality of work for the VDT-based office. But VDTs also have called attention to other factors in the traditional office lighting environment which can have an impact on worker comfort, whether VDTs are in use or not. Since fluorescent lights are the most common type of artificial lighting employed in the office, it is useful to note some other aspects typical of these lights which may affect worker comfort.

Fluorescent lights normally flicker off and on; in fixtures they are installed in pairs, and one light in the pair pulses off as the other is on. The result is that most people cannot perceive any flickering. But if one of the pair burns out or is removed — either for energy savings or as a method of reducing overall luminance—the purpose of this pairing is defeated and noticeable flicker results. Therefore burned-out tubes should be replaced promptly, and if reduced lighting or energy saving is desired, both tubes in a pair should be removed.[30]

Fluorescents also emit different types of light, usually described in terms of warmth. Cool whites are characterized by more bluish light, while warm whites have a more pinkish tone. Cool white is the most common type found in offices. Both cool and warm lights use a frequency range that is substantially narrower than the range found in daylight. The possible biological effects of this type of fluorescent have not been well investigated.[31] Now, however, fluorescents are available which closely simulate the full spectrum range of daylight. While these are more expensive, there are some indications that lighting which approximates natural daylight positively affects people's moods.[32] More controversial theories even speculate that growth, attention span, visual acuity, and endocrine system functioning are improved when full spectrum light is used rather than standard white fluorescents or incandescents.[33] The literature on office lighting is expansive and the reader is referred to the "Further Readings" list for discussions of measurements, guidelines, and techniques.

Eye Examinations and Vision Correction

Because VDT work may increase the demand on the visual system and call attention to uncorrected or undercorrected visual problems, an eye examination may be helpful in preventing visual stress. Many operators discover after an eye exam that they need new prescriptions, or need eyeglasses for the first time. The operator may assume that VDT work "made my eyes worse," but it may be more likely that visual conditions or subtle visual changes went unnoticed until VDT work increased the demand on the visual system. Some studies indicate that 30 percent or more of clerical workers have uncorrected or undercorrected vision.[34]

Operators should inform their optometrists about the nature of their VDT work and the working environment so that the proper eyeglasses can be prescribed. In some cases operators may need a special pair of glasses just for VDT work.[35] In cases of persistent eye problems, it may be advisable for the eye specialist to visit the worksite in person and examine the various components— workstation, job design, environment, and individual vision status.[36] In addition, some individuals who suffer from photosensitive epilepsy may also have special visual concerns about VDT work. These individuals should consult a physician before undertaking VDT work.[37]

Whether all operators should have exams at the time of employment, whether exams should be repeated at specified intervals, and whether the exams and any required eyeglasses ought to be paid for by the employer or by the operator are questions which have more to do with the politics of labor relations and with the move towards standards than they do with the question of visual problems and VDT use. The Newspaper Guild, for example, has asked for paid corrective lenses, annual eye exams, and regulation of lighting for its members who are VDT operators.[38] It seems likely that some employers will make occupational vision benefit plans a part of their benefit packages. The evaluation of a pilot program conducted in New York state for occupational vision care for VDT operators found that the benefit plan under study reduced reports of visual complaints by 93.8 percent and improved the quality of job performance by 82.2 percent among the test group (management/confidential office workers using computers in the governor's office of employee relations). The plan is being continued at the pilot site and has been adopted by several other New York state groups.[39]

Solutions

1. Select monitors which conform to basic guidelines for display quality regarding flicker, resolution, readability and legibility.

2. Select monitors and accessories which provide the necessary adjustments, including contrast, brightness, and tilt and swivel adjustments, and which permit the operator to select either type of display polarity.

3. Make sure that overall lighting levels are appropriate for VDT work.

4. Control glare at the source whenever possible; place VDTs so that they are parallel to direct sources of light such as windows and overhead lights, and use window treatments if necessary.

5. When glare sources cannot be removed, seek appropriate screen treatments such as glare filters. Keep the screen clean.

6. Ensure that the indoor environment—temperature, heat, air flow, and humidity—are within comfortable ranges.

7. Operators should have properly corrected vision, including special glasses for VDT viewing if necessary.

8. Operators should take frequent rest breaks, and glance away

from the screen at distant objects frequently.

9. Operators should make a conscious effort to blink frequently to reduce drying of the eyes.

Conclusions

Even though researchers are still searching for definitions and measurements which will describe and diagnose eyestrain or visual fatigue, and even though the direct contribution of the VDT to these problems may not be measurable until such questions are resolved, several points about VDT work and visual complaints are clear.

There is no convincing evidence that VDTs cause any permanent visual damage, nor is there any convincing evidence that they contribute to the development of cataracts.

A large number of VDT operators do experience symptoms of visual discomfort, but there is disagreement about whether these symptoms are more common among VDT users than among others doing similar work without VDTs.

Most of the symptoms of visual discomfort experienced by VDT operators can be reduced by proper attention to ergonomics and by proper vision correction.

While VDT operators may require changes in vision correction, these changes are not thought to be caused by VDT work. The demanding nature of VDT work may make already existing vision problems more noticeable, and thus more likely to lead to correction. Normal visual changes due to aging may be more noticeable among VDT operators, for the same reason.

Contact lens wearers and bifocal or trifocal wearers may need to make special adjustments in order to perform VDT work comfortably.

Environmental factors which are important to comfort in general, such as temperature, humidity, and air movement, may also affect visual comfort.

Further Readings

Barresi, Barry J. and Rosenthal, Jesse. *New York State Occupational Vision Benefit Plan Study: An Evaluation of a Vision Plan for VDT Users and Office Workers*. State University of New York, Center for Vision Care Policy. 1986.
This report details the methods and results of this pilot study in occupational vision benefits. It is of interest to those who may be considering implementing such vision care plans and also to those who are interested in the impact of visual complaints on worker satisfaction and productivity. It is written in a very readable style and includes several informative tables illustrating the study data.

Boyce, P.R. *Human Factors in Lighting*. Macmillan. 1981.
Boyce, P.R. *Lighting and Visual Display Units*. The Electricity Council Research Center. 1981.
These two resources present lighting requirements in technical terms. *Human Factors in Lighting* explains the visual system and the nature of

light before presenting research and conclusions about lighting requirements. *Lighting and Visual Display Units* is a fifteen-page pamphlet which explains lighting requirements for VDT work areas and presents technical measurements.

Grandjean, Etienne. *Ergonomics in Computerized Offices*. Taylor & Francis. 1987.
This book by a prominent figure in modern ergonomics covers most of the topics which pertain to using VDTs, and it includes chapters on vision, lighting, visual strain, and photometric characteristics of VDTs, and recommendations for workstations. The technical aspects of lighting and of display quality are discussed in readable, although somewhat technical, fashion.

The Koffler Group. "Glare." *Office Systems Ergonomics Report* 4(1) May 1985.
This special issue reviews studies on the effects of glare and on methods of controlling glare, and presents summaries. It discusses different types of glare and how they are measured.

Miller, Stephen C. *The Eye Care Book for Computer Users: A Guide to More Productive and Comfortable Computer Work*. Eye Care Concepts. 1986.
Feeley, Helen. *The VDT Operator's Problem Solver*. Planetary Association for Clean Energy. 1984.
Both of these books are by vision professionals (Miller is an ophthalmologist and Feeley is an optometrist), and both are written for operators. Discussions of how the eyes work, how VDT work can stress them, and what operators can do to help themselves are practical and easy to read. Both books include discussions of vision corrections and other aspects to be aware of when consulting a vision professional.

4   Strain and Injury

- VDT operators report a high frequency of back, neck, shoulder, and general muscle discomfort.

- It is not clear from the studies which have been done whether VDT operators experience these complaints more often than other clerical workers. It is clear, however, that the introduction of VDTs presents new workstation design requirements.

- Prolonged sitting, constrained postures, and poor workstation design are significant contributors to operator discomfort.

- Repetitive motions caused by poor ergonomic design and poor job design may lead to problems requiring medical intervention. Proper attention to these factors can prevent most of these injuries.

- Research indicates that proper workstation design and job design can alleviate most operator complaints of musculoskeletal aches and pains.

- A number of general guidelines for workstation design have been developed to assist in making these changes in the VDT workplace.

Backache is second only to colds and flu as a cause of lost work time in the United States. Estimates indicate that "lower back pain causes about 1,400 lost work days per 1,000 workers every year."[1] Field studies investigating the frequency of musculoskeletal complaints among VDT operators have consistently shown a high incidence of these complaints. In a University of Wisconsin study, 35 percent of the VDT operators reported that they experienced frequent or constant back discomfort and 34 percent reported frequent or constant neck-shoulder discomfort. A NIOSH study reported that 81 percent of VDT operators experienced occasional neck-shoulder discomfort and 78 percent experienced occasional back discomfort. In a large scale NIOSH study 64 percent of the operators reported neck discomfort, 62 percent reported upper back discomfort, and 71 percent reported lower back discomfort with a frequency of "a few times a week to every day."[2]

Strenuous physical work and repetitive factory work have long been known to pose the risk of muscular strain, but office work can also stress the musculoskeletal system and lead to complaints, from the minor to the serious. Neckache and general fatigue are also common complaints among office workers in general. Job-related muscular strains can range from short-term complaints like discomfort and fatigue to medium-term effects such as chronic fatigue or muscular discomfort which leads to medical treatment, and long-term effects such as ailments which lead to ongoing medical treatment. Most complaints are short-term, reversible effects, but these can be precursors of more serious problems if the causes of these complaints are not addressed.[3]

Causes of Muscular Strain

One major component of office work, sitting, is something usually taken for granted. The human body is actually in its healthiest position not while seated, but while standing or lying on one side. In these positions, a slight inward curvature of the lower back (called *lordosis*) is natural. Prolonged sitting stresses the body, particularly the lower back and the thighs, and may cause the lower back (lumbar) region to bow outward if there is inadequate support; this abnormal curvature (called *kyphosis*) can lead to painful lower back problems, a common complaint of workers in all walks of life (see fig. 9).

Much office work involves not only prolonged sitting, but sitting while using the hands, arms, or eyes to perform tasks which are essentially stationary. Thus muscles are being used, but are not being flexed between a state of use and a state of relaxation. When the muscles of the arms, neck, shoulders, legs, and hands are in a fairly continual state of contraction, a condition known as static loading occurs. In this condition, because the muscles are contracted but not rhythmically relaxed, circulation is decreased and waste products like lactic acid are not carried away from the muscles as rapidly as they should be. Also owing to decreased circulation, nutrients carried in the blood are not supplied in sufficient amounts. Too many waste products and too little nourishment combine to produce muscle soreness. This is the most common cause of muscular complaint. Although it can be quite painful, it does not cause true injury to the muscles since increased blood flow through rhythmic contraction and relaxation soon washes away the waste products and nourishes the muscles.[4]

Neckache and other discomfort can also result from mechanical wear and fatigue caused by excessive movement, such as turning or twisting the head frequently from side to side or frequent tilting up and down. Placement of work materials is often at fault in these instances. Poor posture, such as slouching, can also lead to postural discomfort and is often the result of poor seating or workstation design. Individual differences such as obesity or pregnancy (both of which may compress nerves and decrease circulation to the legs) and habits such as leg-crossing may also cause discomfort.[5]

Even though most of these complaints do not lead to actual injury to the worker, they do result in a significant loss in worktime, productivity, and worker health, and as such they represent an occupational health problem. In some cases, chronic injury can result from repetitive work, and from non-work activities as well, such as some types of recreation. The most prevalent of these is repetitive strain injury, or RSI (also called repetition stress injury).

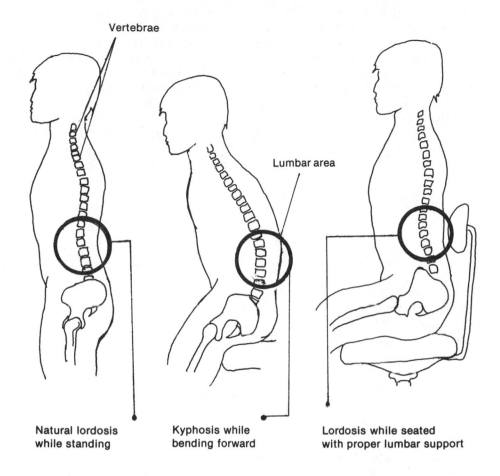

Vertebrae

Lumbar area

Natural lordosis
while standing

Kyphosis while
bending forward

Lordosis while seated
with proper lumbar support

Fig. 9. The Lumbar Curve in Three Common Positions

Repetitive Strain Injury (RSI)

Unrelieved repetitive motion has been identified in many activities as the cause of localized injury, such as bowler's thumb, tennis elbow, and washerwoman's thumb, all forms of RSI. In some hand-intensive occupations as much as 25 percent of the workforce may have these repetitive motion disorders of the hand and wrist.[6]

While RSI can refer to chronic pain in muscle groups, single muscles, or large areas, it is most commonly used to describe specific syndromes involving the nerves, muscles, and tendons, particularly of the upper body.[7] Ailments which fall into this category include tenosynovitis, tendinitis, rotator cuff syndrome, and carpal tunnel syndrome. It is commonly associated with "some pathological abnormality such as degeneration, inflammation, calcification, microfracture, etc."[8]

Occupational RSI usually develops over time, beginning with symptoms that occur intermittently and which are relieved by a period of rest. As the individual continues to work the symptoms may increase in frequency and intensity until measurable changes such as swelling appear, changes which are not relieved by normal time away from work. The physiological mechanisms which cause RSI are not well understood, but may include local oxygen depletion, lack of lubricating fluid to the tendons, toxin build-up, and changes in metabolism.

RSI treatment can include change of work, rest, joint immobilization, drugs, physical therapy, occupational therapy, surgery, instruction in various stress management techniques, counseling, and acupuncture or chiropractic treatments. Whole body exercise may also help in both the treatment and prevention of RSI. The efficacy of the various treatments is not well documented.[9]

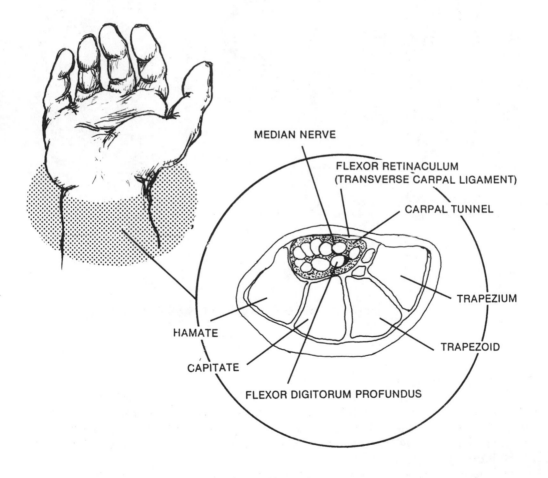

Fig. 10. The Hand and Wrist with Cross-Section

MEDIAN NERVE

FLEXOR RETINACULUM
(TRANSVERSE CARPAL LIGAMENT)

CARPAL TUNNEL

TRAPEZIUM

TRAPEZOID

FLEXOR DIGITORUM PROFUNDUS

CAPITATE

HAMATE

One of the most serious forms of RSI is carpal tunnel syndrome. Carpal tunnel syndrome is seen in occupations such as carpentry which place high demands on the wrist-hand system. It has also been reported among VDT workers. The carpal tunnel is a narrow passageway formed on the top by the wrist bones and on the bottom by the transverse carpal ligament. Through this tunnel pass the tendons from the forearm and the blood vessels to the hand (see fig. 10). The median nerve lies against the transverse carpal ligament. Constriction of this tunnel through repeated flexion (upward bending) or repeated extension (downward bending) of the wrists, or a combination of both, can compress and damage the median nerve.

Carpal tunnel syndrome may be characterized by numbness or tingling in the hand, especially the first three fingers, or by weakness and loss of gripping control. Often the first symptoms are burning, aching night pain. Later stages of the disease may produce paralysis and atrophy of the thumb muscles. It can be very effectively treated with surgery.[10] It is much more common in women than men, particularly menopausal women or those who are pregnant, so hormonal changes may play a role in its development.[11] It is also likely that heredity is a factor.

VDT Work and Muscular Strain

Backache, Neckache, and Sore Muscles
The typical seated office worker, and even more so the seated VDT operator, spends most of his or her time in a stationary position. Many of the tasks which were once done sitting at a typewriter or desk are now done sitting at a computer terminal. Sitting at a typewriter or a desk for eight hours a day, five days a week, can easily lead to sore, tired muscles, the familiar "I want to walk around, I've been sitting all day" feeling. Add to prolonged sitting the relatively rigid position of the head and neck when reading or typing, and it is easy to see how static loading can lead to neck and back pain.

The VDT adds some new factors to the office environment. The terminal is not a piece of paper. The screen faces the operator, presenting the work in a vertical plane instead of the horizontal plane of the traditional desk. This means that the head and neck must adopt a position that is new to seated work, one that will almost certainly be uncomfortable if the terminal and keyboard have simply been laid out on a desk designed for paper-based tasks.

The VDT keyboard also introduces its own problems. Older model terminals may have the keyboard and screen in one case, making it impossible to alter the distance between screen, keyboard, and user, or to raise or lower either screen or keyboard to more comfortable heights. Workers tied to a VDT are also less likely to exchange comments with others, get up to get files, take notes, and in general divert their attention and change their posture.

Most researchers suspect that the high incidence of back and neck pain reported by VDT operators is a reflection of the nature of the task (repetitive clerical work with static postures) rather than the tool (VDTs, typewriters, or paper). According to Thomas Läubli, "The groups with most complaints are characterized by repetitive tasks combined with lasting and uninterrupted sitting or standing. Frequent armpains are found in groups with a high number of daily keystrokes." His studies further revealed "that the ergonomic shortcomings [of VDT workstations] have clear effects on postural adaptation and that skeletomuscular disorders are increased in the respective parts of the body."[12] Thus task design and workstation design both play important roles in the development or prevention of muscular strain in general, and RSI in particular.

VDT Work and Repetitive Strain Injury (RSI)

Reaching for typing paper, rolling it in and out of the typewriter, manually filing the completed documents, and all the minute actions of the paper-based office provide some change of position, no matter how small these actions seem. But many VDT operators need only to push keys and perhaps turn the pages of a source document. This intensifies what is already a fairly static task. The combination of repetitive action, static loading, and concentrated mental and visual effort can lead to generalized muscular discomfort, "one of the most common subjective complaints among VDU operators."[13] If this repetitive motion (entering keystrokes) is uninterrupted, if the arms, hands, and wrists are forced into uncomfortable positions, and if there is pressure to increase the number of keystrokes per minute, the operator may experience strain in the arms, hands, or wrists. Over time, this strain could result in cumulative trauma such as RSI. The VDT task may particularly stress the wrists.

The suspected causes of RSI among VDT workers have been investigated most thoroughly in Australia, where, according to one source, over 4,000 cases of RSI among VDT operators have been reported in the few years since VDTs were introduced there. Eighty percent of these cases are keyboard or word processing operators.[14] Investigations of the Australian experience emphasize the impact of reduced task diversity, increased workload, and other adverse changes in job design brought about by the introduction of VDTs on the incidence of RSI.[15]

Poor workplace design in general is also a factor. Australia has introduced VDTs very rapidly, sometimes bypassing the transition stages (such as electronic, or even electric, typewriters), and also has done so with little planning for VDT workplaces. Some observers suggest that as a result VDTs have been a greater shock to Australian keyboard users than to those in other Western countries. Three Sydney daily newspapers rushed to introduce VDTs into newsrooms, where they "were simply squeezed in among existing desks. Wiring hangs in tubes from the ceiling. Reporters complain that the noisy, cramped conditions and poorly designed furniture add to stress and fatigue." The Sydney dailies have experienced a high incidence of RSI cases.[16]

There is, however, also some dispute about the diagnosis of RSI in Australia. RSI is distinct from the fatigue of the neck, back, and legs often reported in VDT operators. RSI is not merely discomfort which is relieved by rest, but is in fact an injury. It is frequently debilitating. Australian RSI figures, however, apparently include complaints that would elsewhere be classified as short- or medium-term discomfort. Some researchers suggest that "muscular fatigue related to working conditions" better describes cases which do not appear to involve actual injury.[17]

Australian labor unions are particularly active, and union concern about RSI may have made workers there more aware of such complaints, resulting in greater reporting of these ailments than would otherwise have occurred. Another factor which may have led to more frequent reporting is the workers' compensation system in Australia, which recognizes more types of RSI complaints than the compensation system in, for example, the United States. Finally, Australia's record keeping on RSI seems to be particularly complete.[18]

There may be many reasons for the apparently greater frequency of RSI in Australia; however, it is clear that RSI among VDT operators is now being recognized more often in the United States as well. Twenty-two cases of RSI have been reported in the newsroom at the *Los Angeles Times*, where employees have started calling it "computeritis." Four cases of carpal tunnel syndrome were reported in 1985 to the San Jose Newspaper Guild, although none had been reported to the union previous to the introduction of VDTs at the local paper some twelve years ago.[19] Other cases have been reported at the *Philadelphia Inquirer*, the *San Diego Tribune*, the *San Francisco Chronicle*, the *San Francisco Examiner*, *Newsday*, and the *New York Daily News*.[20]

Measuring Muscular Discomfort

Exactly how much discomfort is directly related to VDT work is disputed. According to researcher Robert Arndt,

> While several researchers have reported a higher incidence of problems among VDT users, others have found very few differences. Other studies have found that the frequency and type of problems are more closely related to the nature of the VDT task than to the use of VDT's *per se* . . . . While it is clear that postural problems and musculoskeletal complaints are not limited to VDT work, it can be concluded that the unique characteristics of VDT equipment, workstations and tasks must be considered as additional contributing factors.[21]

Arndt specifically mentions such contributing factors as the vertical orientation of the screen, limited equipment mobility, increased space needs, and the tendency to simplify tasks and make them more repetitive when VDTs are introduced.[22]

Other researchers, particularly Martin Helander and his colleagues, have criticized the methods of many of the field studies.[23] According to them, one main drawback of many studies has been the questionable comparability of the control groups of non-VDT workers in studies investigating whether VDT operators have significantly more complaints of this type than non-VDT operators doing the same work. Most research on postural problems and muscular discomfort among VDT workers has been in the form of field studies (although a few well-documented and carefully controlled laboratory experiments have been conducted). As with field studies examining vision and VDT work, many studies have used control groups of non-VDT users whose work was not directly comparable to that of the VDT user group (see chap. 3 for discussion of control groups). As with vision studies, questionnaires and surveys are often used. Problems with control group selection and problems with questionnaires make many of these studies unreliable, and the results are often conflicting. While the answer to the question "Do VDT workers suffer more postural and muscular discomfort than non-VDT users?" is certainly important, the most relevant fact is that both VDT and non-VDT operators performing clerical work report a high incidence of back, neck, shoulder, and arm pain. This constitutes an occupational health problem that is largely avoidable, according to most ergonomists.

Most experts believe that appropriate workstation design and task design can alleviate the major portion of work-related muscular strain complaints.[24]

Researchers in ergonomics have studied VDT operators at their workplaces and in special simulated workstations in laboratories to learn what changes can be made in the workplace to make VDT work more comfortable and less stressful. Some of the studies have been observational— working postures were simply observed and recorded, using measurements, videotapes, and/or photographs. Frequently, operators are asked to report discomforts and/or fill out surveys or questionnaires.[25] Other studies have been interventional—some change was made in the workplace, such as the introduction of a new chair or workstation, and the changes in posture and in reported complaints were recorded. (Interventional studies can be unreliable unless efforts are made to control or correct for the Hawthorne effect, in which the very fact that a change has been made is likely to produce positive results.[26])

These studies have led to a number of proposals for changes in the workstation, the equipment, and the work environment. Earlier recommendations based on measurements of body size and body mechanics using techniques in anthropometrics and biomechanics were often found to be different from conclusions drawn from field studies. However, the uneven quality of the research makes it difficult to know whether the findings can be trusted. As a result, recommendations for such values as chair and worksurface height vary in the many standards and guidelines available.

Reducing Muscular Strain in VDT Work

In spite of the diversity of opinion about specific measurements, there are several points about which most authorities agree. This section will mention measurements only in passing, and will instead focus on general workstation requirements which should be considered when VDT equipment is introduced into the office. A list of further readings at the end of the chapter will lead the interested reader to detailed discussions of guidelines and studies.

Chairs

The office chair is far from simple. A chair that is improperly designed for the task or not adjustable to the individual can restrict blood flow to the lower legs, provide inadequate support to the back, or force the user to maintain a static position for extended periods.[27] It is ironic that airlines and car manufacturers advertise wider seats with more legroom as luxury features while the average office worker is seldom aware of the importance of good design in office chairs. Instead, the size of the chair and the quality of the chair covering have often been symbols of office status rather than worker comfort.[28] Only recently have chair advertisements regularly emphasized ergonomic design features, even though "typical office workers spend more time in their chairs than in any other piece of furniture except their own beds."[29]

Conflicting findings have made it difficult to determine the right adjustability for a VDT operator's chair. The eminent Swiss ergonomist Etienne Grandjean noticed that many operators prefer a backward-leaning posture,[30] while Danish surgeon A.C. Mandal recommends that chairs be designed to tilt forward to retain the lumbar curve (lordosis).[31]

The differences of opinion may be due to differences in tasks being performed. Different tasks may lend themselves to different operator positions. The demands of intensive data entry are different from those of verification or

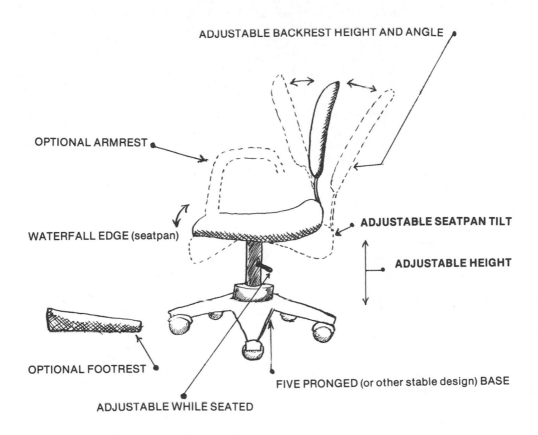

ADJUSTABLE BACKREST HEIGHT AND ANGLE

OPTIONAL ARMREST

WATERFALL EDGE (seatpan)

ADJUSTABLE SEATPAN TILT

ADJUSTABLE HEIGHT

OPTIONAL FOOTREST

FIVE PRONGED (or other stable design) BASE

ADJUSTABLE WHILE SEATED

Fig. 11. Some Features of An Adjustable Chair

proofreading of entry, or of programming. Thus to accommodate these different tasks an ergonomically sound chair requires four degrees of freedom—seat pan tilt, backrest angle, seat height, and backrest height. Operators can then vary the chair adjustments according to the task. One recent study indicated that operators do in fact make the expected adjustments when informed about the basics of spinal anatomy and muscular fatigue and properly trained in the use of the chairs.[32]

While there are differences of opinion on some fine points, there is general agreement about the following principles (see fig. 11):

1. The chair must provide height adjustability.

     Because people are of different heights (and people of the same height have different proportions) the adjustability range must be sufficient to accommodate a wide range of users, and it must be easy to make these adjustments while seated. Gas or hydraulic lift mechanisms are commonly used on better chairs.[33]

2. The seat edge must not press against the underside of the thigh.

     This constricts the flow of blood to the legs. In practice this means that the seat pan must not be too long from front to back, and that it must be padded and have no sharp edge. The edge of the seat pan should curve away from the legs, in what is called a "waterfall edge."

3. Backrests should have height adjustments and provide lumbar support.

     Support for the lumbar area is crucial, as poor lumbar support is a major factor in lower back pain (the proposed ANSI standard requires backrests with lumbar support, but does not require them to be adjustable).[34]

4. The chair should allow freedom of motion to allow the user to change position easily and frequently without losing good support.

     This generally means that seat pans and/or back rests must have some "give" when pressure is exerted. Users should also be able to lock the seat pan in various positions to allow for backward-leaning postures as well as forward-sloping ones, according to the task and the individual's preferences.

5. The base of the chair should provide stability and protection against tipping.

     This usually means that the chair base will have a five-star design instead of a four-star base, although other designs which are equally stable could be used.

6. Optional armrests may be used.

     Armrests should not interfere with the ability to move the chair under the worktable.

Workstations

For certain types of seated work, such as drafting, the traditional desk has long been considered inadequate. Now it is generally believed that the rigid, horizontal desk is not always adequate for work at VDTs, either. There is, however, no general agreement about what is appropriate. Recommended heights for keyboards and monitors vary greatly, and observed preferences of operators encompass a wide range. Thomas Läubli sums up the difficulty:

Fig. 12. Workstation with Proposed ANSI Measurements

Imposed dimensions—even if they are similar to the mean values of preferred settings—cause significantly more postural discomfort. It is concluded that adjustable VDT workstations may be properly fixed by the individual operators, guided by their feelings of relaxed posture. There is no recommended single uniform dimension that can suit everybody.[35]

But adjustability must be within the realm of good ergonomics; as Tom Stewart cautions, "Making everything adjustable is no substitute for good design."[36]

Thus the only firm recommendation that can be made is that adjustability is required in many of the components of the workstation. Desirable attributes of a workstation would then include the following:

1. Height, tilt and swivel, and fore-aft adjustments for the monitor support.

2. Sufficient flat workspace area for source documents and other materials.

3. In some cases, a separate keyboard platform which is adjustable for height and tilt and for fore-aft settings.

If a thin keyboard is used, it may not be necessary to provide a separate keyboard support. In practice, an adjustable, single-level surface combined with an adjustable pedestal for the terminal and a thin keyboard might be sufficient for a VDT workstation. When only one worker uses a workstation, it is adequate to have the worksurface adjusted to a fixed height, so long as the rest of the workstation meets the above guidelines. As in any traditional office setting, basic principles of ergonomics must be followed, such as the provision of ample leg room and a well-designed adjustable chair.

The workstation does not exist by itself; workstation, operator, and chair are all part of a system. This helps to explain the variety observed in operator postures. When the operator is in the forward-inclined posture for certain tasks, he or she may wish to adjust the levels of the workstation quite differently from the levels that are comfortable when leaning back for other tasks.[37] For example, Grandjean, Hünting and Pidermann observed that users tend to set the keyboard much higher than was expected when this backward-leaning posture is adopted.[38]

There are recommendations and standards which present specific ranges of adjustment, and while they differ from one source to the next, the basic ranges tend to be comparable. In figure 12, the dimensions are taken from the proposed ANSI standard which has been circulated for review and comment. It will, if officially accepted, be adopted as the voluntary standard for equipment manufacturers in the United States.[39]

Keyboards

As discussed above, excessive strain to the hands, arms, and wrists can lead to some of the more serious chronic problems that have been associated with VDT work. There are several things which can be done to reduce this strain. "The most desirable workstation layout for minimizing the risk of a repetitive trauma disorder is one in which the task can be performed with the elbow at the side of the body, without excessive forearm rotation, and without the worker deviating, flexing, or fully extending the wrist."[40] A wristrest which allows the wrist and hand to assume a more relaxed position and a keyboard height which permits the forearms, hands, and wrists to adopt a position more or less parallel to the floor are recommended.[41] It is also important for the operator to take frequent brief breaks, moving at least the arms, hands and wrists, but preferably the whole body. Interspersing non-VDT tasks with VDT tasks is strongly recommended (task alternation and rest breaks are discussed further in chap. 5).

Certain design features of the keyboard can reduce muscular strain and affect other aspects of the VDT environment such as reflected glare. The following points should be kept in mind when selecting a keyboard.

1. For most office tasks, ergonomists agree that the keyboard should be detached from the display screen so

that it can be easily moved. Not every task is the same, however, and for some types of tasks, such as brief inquiries made at walk-up terminals, detachable keyboards may not be necessary or even desirable since they take up more space.

2. Keyboards should have a matte finish to reduce reflections.

3. Most ergonomists also recommend a slim keyboard (to reduce extension of the hand which can increase pressure on the carpal tunnel) that can be adjusted in the degree of tilt. This is often provided by the addition of a separate, adjustable keyboard holder.

4. The keys should be slightly concave in shape and should have a light but distinct response. Some form of feedback, either tactile (a distinct feel that a key has been activated) or audible (such as a keyclick), should be provided. If audible feedback is used the operator should have the option of turning it off.

5. Most authorities also recommend wristrests, although there is a wide range of opinion on the best design.[42]

There is also a large body of research on the actual design of the keyboard layout, because of the additional function keys and numeric keypads which are part of most computer keyboards. New layouts for the basic alphanumeric keys have also been proposed. The Dvorak keyboard, for example, is arranged so that the most frequently used keys are also the most conveniently located. However, this and other alternatives have not been widely accepted in the marketplace, and the familiar QWERTY keyboard (so named because of the arrangement of the first five keys on the row directly above the home, or resting, row) remains the accepted standard. Several layouts for the numeric keypad are acceptable. Some ergonomists (notably Etienne Grandjean and K.H.E. Kroemer) are experimenting with alternate keyboard designs such as the two-piece "split" keyboard.[43]

Accessories

Footrests. Because even the most adjustable chairs and workstations may not have sufficient range to accommodate everyone, some users may require a footrest. An inclined footrest with a large surface area, preferably with adjustable height, should be provided. It should not, however, restrict the under-table leg space too greatly.

Document Holders. For tasks which rely on paper source documents (hardcopy), a suitable document holder should be provided. Several authorities recommend that the copy be at or near the height of the screen, but in some applications it can be placed between the screen and the keyboard. In still other cases, the copy might be the primary focus, and the monitor might be off to the side. Again, operator preference and task requirements will have a great deal to do with the placement of copy, thus maximum adjustability is desirable.

It is clear that copy which is placed flat on a horizontal plane (such as a desk) is not appropriate for intensive data entry tasks. A study by Jan-Erik Hansson and Monika A. Attebrant concluded that "the height of the visual

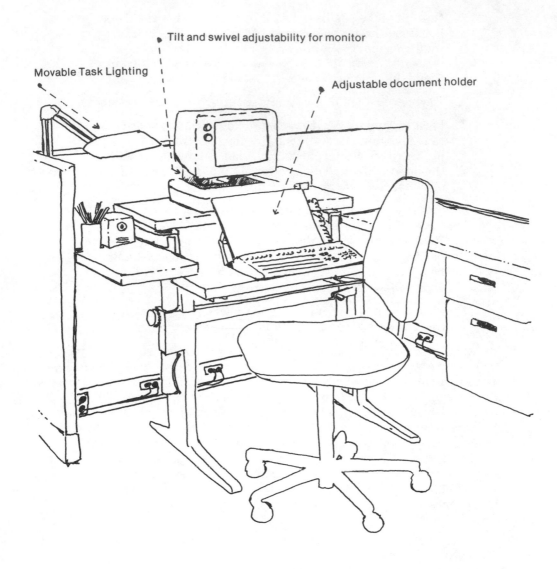

Tilt and swivel adjustability for monitor

Movable Task Lighting

Adjustable document holder

Fig. 13. Workstation Accessories That Can Improve Postural and Visual Conditions

object (manuscript, etc.) is of major importance to the position of the head."[44] Properly placed copy will help prevent uncomfortable postures and will also reduce excessive head and eye movement.

Task Lighting. To illuminate the copy, specialized lighting may also be desirable.

Figure 13 shows how some common accessories might be arranged at a workstation.

Operator Interventions

VDT operators themselves can contribute to greater comfort at the workstation by taking the time to use adjustments which are provided and by using breaks to take the opportunity to move around or even perform simple exercises to help increase circulation and reduce muscle tension. Several pamphlets and brochures are available with suggested exercises, and at the end of this chapter some sources are cited. General physical well-being, promoted by sound exercise habits and nutrition, is an important partner to good workstation design in reducing muscular strain and injury.

The Workstation Environment

The recent focus on office ergonomics has in fact centered on workstation design, chair design, keyboards, and glare treatments—in other words, factors which seemed to contribute directly to postural and visual comfort. But VDT work involves much more than an operator, a terminal, and a workstation. A survey of nearly 4,000 British VDT operators indicated that providing equipment designed for the VDT such as adjustable workstations is not effective if the overall environment— including noise level, light, and heat—is considered to be unsatisfactory. The survey concludes: "It appears, therefore, that there is a complex interaction between the physical, ergonomic and socio-psychological factors associated with the introduction of new technology. All these aspects of VDU use should be taken into account when searching for solutions to the health problems of VDU operators."[45]

The environmental requirements of VDT operators are not unique; they are essentially the same as the needs of other office workers and of people in general. The discussions which follow are thus nearly as applicable to offices in general as they are to the automated office in particular. The VDT does contribute some new elements to the office environment, however, and these elements will be the focus of the discussions.

Indoor Temperature

Temperature preferences vary greatly among individuals, and the ambient room temperature is almost never satisfactory to everyone. However, in general, a room that is too warm makes people feel tired, and one that is too cold causes people's attention to drift and makes them restless. Temperature is not constant throughout an office environment, due to such factors as glass, heat sources (VDTs, people, and others), and placement of climate control devices. However, a general recommendation is that the temperature be held as constant as possible in the range of 21 to 23°C (70-75°F), and slightly warmer in the summer (so as to minimize the discrepancy between indoor and outdoor temperature), in the range of 26-28°C (78-82°F).[46]

VDTs can generate from 100 to 400 watts of heat, compared to the 100 watts or so produced by a human body. It is thus important to calculate the thermal load and to consider the air handling system when introducing VDTs. The dry conditions promoted by the extra heat should also be taken into account, since dry air promotes static electricity. VDTs may increase the thermal loading in an office by as much as 30 to 150 percent.

By choosing VDTs with low heat emissions, venting them away from operators, and distributing them throughout the area, the increased heat load can be controlled somewhat. This is preferable to trying to control the heat load through additional air conditioning efforts, because drafts and other aspects of air conditioning are often judged to be uncomfortable.[47]

Indoor Humidity

There are powerful reasons for introducing measures to keep the office climate at about 50 percent relative humidity.[48] People are generally more comfortable in this humidity range, and workers may find that they have fewer winter respiratory problems and that their moods are generally improved.[49] Care should be taken, however, to maintain a relative humidity below 70 percent since high humidity is associated with discomfort and can contribute to the development of bacterial and fungal growth, especially in sealed (or "tight") buildings.[50]

The moisture content of the air can be increased by adding water to the incoming air via either humidifiers or air washers. Another method of raising the relative humidity is the introduction of plants. Plants provide other benefits as well, since they give off oxygen and remove carbon dioxide from the air, and also provide a pleasant visual diversion.

The increased thermal load generated by the introduction of VDTs into the office environment can lead to drier air, which, in turn, promotes the development of static electricity. The electronic circuits in VDTs, central processing units, and printers are particularly sensitive to static electricity, and loss of data and other problems can result. Raising the relative humidity is highly effective in reducing static. In the electronic office, there is another reason for concern about static electricity, since static fields around VDTs have been implicated in certain types of facial rash among VDT operators.

VDTs and Skin Rashes. A number of cases of skin rash, primarily facial rash, have been reported among VDT operators. Most of these reports are from Norway and Sweden, with a few from the United Kingdom. Researchers have speculated that excessive static (which is promoted by dry conditions that are especially common in cold climates) may be responsible. Heat generated by VDTs can dry the air. One explanation for the reported facial rashes is that charged particles in the air—dust, spores, and other suspended particulates— are attracted to the static field set up by the VDT.[51] These particles may then bombard the operator, and individuals who are sensitive may experience a rash reaction.[52] It was also noted that the carpeting in offices where the rashes occurred was frequently of a type that is susceptible to electrostatic charging.[53] It may be that sealed windows and lower air recirculation rates which are commonly used to conserve energy in so-called "tight buildings" play a role here, too, since they may lead to higher concentrations of particulates, which can include residual contaminants from smoking and other sources.

Body potentials (electrical charges) of 5,000 to 10,000 volts are not uncommon in environments with high static fields, and some studies have noted that operators affected by rashes have carried unusually high charges.[54] A causal relationship between any one aspect of VDT work and facial rashes has been difficult to establish, partly because the reports of skin rash have thus far been localized, and partly because of the difficulties involved in measuring static fields in the office environment. Reporting of rashes has also decreased since the initial outbreaks were reported, possibly because radiation— which was suspected at first—has been refuted as a cause, thus alleviating operator anxiety. However, in published reports, many of the rash cases improved on non-working days, hence a relationship between the rashes and VDT work may be indicated.[55] Although research in this area is continuing, simply increasing the relative humidity in the office environment to at least 30 percent and preferably around 50 percent is likely to have many benefits, regardless of any correlation which might be established between charged particles and facial rashes.[56] Grounding the screen with anti-static filters might be helpful only if a relation between electrostatic fields is established.[57]

Solutions

1. Provide a well-designed adjustable chair, a properly designed workstation (with adjustability where recommended), and accessories such as document holder, task lighting, and footrest where needed.

2. Design jobs to incorporate varied tasks in order to reduce repetitive components and increase freedom of motion.

3. Provide training in the use of adjustable components and encourage their use not only at the beginning of work, but also throughout the workday as tasks and body position change.

4. Allow frequent rest breaks, and encourage operators to use them to move around. Breaks may be combined with alternate tasks to reduce the amount of sedentary work.

5. Compensate for the extra heat load from VDTs by paying special attention to the indoor environment. Maintain a comfortable temperature with as few drafts as possible. Keep the relative humidity at around 50 percent to reduce static, improve comfort, and improve worker resistance to respiratory infections and skin rashes which may be static-related.

6. Operators should avoid excessive extension or flexing of the wrists. Wristrests or keyboard slant adjustments can be useful in this.

7. Operators should avoid excessive turning, twisting, and tilting. Rearrange work materials to reduce the necessity of adopting unnatural postures, if necessary.

8. Operators should also be aware that a personal exercise program to improve muscle tone and circulation can also help to increase well-being and comfort at work.

Conclusions

Backache, neckache, and general muscle pain and fatigue are common complaints of all workers, not just of VDT workers. The incidence of these complaints among VDT workers is high, but it is not clear whether it is any higher than for other workers performing similar tasks. Most of these complaints are reversible, short-term conditions, but chronic pains leading to medical treatment and even some injuries requiring medical treatment and/or surgery can result from certain types of work, including VDT work.

However, the incidence of these complaints can be reduced greatly by improved workstation design, improved placement of work materials, and more varied tasks. While the specific recommendations for workstation design are often based either on anthropometric or biomechanical models that may not be relevant, or on research results that are contradictory, there is general expert agreement on certain basic concepts of workstation design.

Improvements to the working environment should not, however, focus shortsightedly on the workstation alone. Thermal comfort and humidity are also important parts of a good ergonomic environment. In particular, humidity may help reduce certain types of occupational dermatitis and can help prevent static-related hardware problems. Overall worker health and habits can also make a positive contribution to comfort.

Further Readings

Dainoff, Marvin J. and Dainoff, Marilyn Hecht. *People & Productivity: A Manager's Guide to Ergonomics in the Electronic Office*. Holt, Rinehart, & Winston of Canada. 1986.
This book is written for the non-specialist and presents very readable discussions of seating, workstation design, and workplace environment and design. It also addresses vision and display quality, software ergonomics, job design, and the evaluation of scientific research. Of particular interest are discussions of how to evaluate ergonomic chairs and a case study called "Putting Together a VDT Workstation."

Donkin, Scott W. *Sitting on the Job: A Practical Survival Guide for People Who Earn Their Living While Sitting*. Parallel Integration. 1987.
Donkin, a chiropractor, has written a thorough manual which emphasizes that both employers and employees must work together in a team effort to reduce physical discomfort at work. Discussions of basic body mechanics and personal behaviors emphasize the operator's role in comfort; examinations of the workspace, workstation, materials, and environment aid operators in identifying trouble spots. The various physical complaints common to operators are discussed individually, and separate chapters on stress, sleep (including healthy ways to sleep), exercise, and habits wrap up this very readable, systems-based approach. It includes many useful illustrations. An interesting aspect of this book is that it discusses all office work and then makes specific treatment of VDT workplaces.

Knave, Bengt, and Widebäck, Per-Gunnar, eds. *Work with Display Units 86.* North-Holland. 1987.

These are the selected edited proceedings of the Work with Display Units conference, and they include about one hundred of the more than three hundred papers which were presented in Stockholm. Virtually every area of concern regarding VDT work is covered. Sections on Working Posture, Workplace Design, Physical Inactivity, and Skin will be particularly pertinent to the subjects discussed in this chapter. Many other sections, such as Work Organization, Information and Education, and Task and Stress, contain papers which are also interrelated with musculoskeletal issues. These papers will provide a good introduction to the more specialized literature on VDT work without being too technical for the interested lay reader.

The Koffler Group. *Office Systems Ergonomics Report.*

This newletter reports on research and each issue focuses on a special topic. These have included issues on keyboards, physical and environmental factors, glare, and research methodology and evaluation. The newsletter has monitored the ongoing VDT health debate very closely.

Lueder, Rani, ed. *The Ergonomics Payoff: Designing the Electronic Office.* Holt, Rinehart & Winston of Canada. 1986.

This is a useful resource for those engaged in office space planning and design and in the purchase and specification of office furniture and equipment. It features nuts and bolts discussions of workstation design, wiring and cabling, environmental comfort, and display/lighting requirements.

5   Stress

- Stress-related diseases such as coronary heart disease, psychological disorders, and ulcers, along with stress-linked actions such as accidents, substance abuse, and absenteeism cost U.S. industry as much as $150 billion dollars a year.

- Stress is the body's response to the physical, emotional, and environmental demands which are made upon it.

- When the body does not have the time or resources to recover from daily stress, chronic stress results.

- Chronic stress can deplete the body's resources and can lead to such stress-related diseases as coronary heart disease, high blood pressure, ulcers, psychological disorders, and substance abuse.

- It has recently been found that clerical workers in general experience high levels of job-related stress, higher than other groups, such as executives, who were formerly thought to have high stress levels.

- VDT operators may experience stress because of job insecurity, negative feelings about computer monitoring, shiftwork, machine pacing, incentive pay schemes, increased workload, repetitive work, role ambiguities, reduced social interaction, and lack of opportunities for advancement.

- Careful attention to job design and training can eliminate or alleviate most of the problems associated with VDT work.

All humans experience stress. While there is no precise definition of stress which is universally accepted, it may generally be described as the result of any demand upon the body.[1] Stress is inevitable, and it is not always negative. In fact, responding successfully to some types of stress such as appropriate challenges (in work or in recreation) can provide an important source of personal satisfaction. However, if the stress response is in a continual state of activation, the body's emergency resources become depleted, reducing the immune system's ability to fight infection. Chronic stress may also lead to the breakdown of organ systems and to specific diseases which are directly related to the physiological mechanisms involved in the stress response.

The Costs of Stress

A 1985 Office of Technology Assessment study, *Automation of America's Offices*, concluded that "stress-related illness or disease . . . may emerge as the greatest public health problem among office workers in the future."[2] Mental health and stress-related conditions account for 40 million lost work days per year, according to the National Association of Mental Health.[3] The cost to U.S. industry for stress-linked disease (coronary heart disease, psychological disorders, ulcers, accidents, substance abuse, absenteeism) may be "as high as $150 billion dollars a year."[4] Medical claims for work-related psychological disturbances such as neuroses and for mental stress which develops gradually over time are increasing at rates which represent "substantial health and financial costs in the United States." The evidence indicates that "an unsatisfactory work environment may contribute to psychological disorders."[5]

This cost may be even higher as the trend continues for courts to find that "*companies are responsible* for the physical and mental well-being of the people working for them" (emphasis in original).[6] Management may also pay for stress in increased labor costs and decreased organizational control as unions begin to focus on job and organizational design in an attempt to address the issue of worker stress.[7]

In the controversy over whether or not VDT work can lead to reproductive problems, the question of stress has taken on new importance. Two researchers, Ericson and Källen, claim that increased risk of reproductive problems (particularly low birth weight, prematurity, and perinatal death), if present, is "most likely an effect of co-varying risk factors like stress and smoking."[8] Stress seems to result when an individual perceives that something requires an adaptive response, regardless of whether or not such a coping reaction is actually required. That is, perceived danger may be as stressful as actual danger. If this is true, "whether significant health problems are associated with VDT use or not is not of consequence to stress considerations; employees perceive that they are and are fearful for their health. They are stressed by this fear."[9] Some authors have suggested that such anxiety itself may be a cause of reproductive difficulty among operators.[10]

Stress and Strain

In earlier chapters various visual and muscular complaints common among VDT users were discussed. It is important to realize that these complaints are also the result of stress. Something (such as poor chair design) stresses a system in the body, and the body tries to cope with the stressor (by changing posture, for example). If the individual cannot compensate for the stressor adequately, pain or injury—*strain*— may result. While strains are not diseases as such, "they can be regarded as risk factors for physical and mental illnesses."[11] The relatively concrete results of stress leading to strain—visual or postural discomfort—are, to some extent, easier to recognize than the effects of mental or psychosocial strain and may seem somehow more legitimate. Backaches are a leading cause of absenteeism from work, but equally important strains due to mental stressors may not receive attention. If environmental conditions such as temperature, noise, and air quality are poor, they can also be powerful stressors which may easily go unrecognized.[12]

The consequences of continued stress—mental or physical— can be quite serious. Stress is known to play an important role in coronary heart disease, coronary artery disease, high blood pressure, and ulcers. Increased reliance on coping mechanisms like cigarettes, alcohol, or drugs may increase the health risks associated with excessive smoking and substance abuse. Less severe effects of chronic stress include aches and pains, fatigue, sleep and eating disruptions, and behavioral/attitudinal changes.

In order to understand the complex ways in which stress, especially chronic stress, can affect health, it is useful to understand what the body does when it experiences stress.

How the Body Reacts to Stress

When the human body experiences stress—whether emotional or physical— a complex set of responses is available to attempt to cope with the stressor(s). Hans Selye's influential physiological model of the generalized stress response, also called the general adaptation syndrome, describes the physiological reactions which may result from stress. These reactions are designed to increase energy production and help the organ systems to operate at high efficiency. Not every stressor will evoke all of the body's possible responses. Recent stress research has, in fact, begun to look at specific as well as generalized (nonspecific) responses.[13]

The earliest reaction to stress is the release of hormones which then stimulate many of the basic stress mechanisms. The pituitary gland produces ACTH, a hormone which stimulates the secretion of epinephrine (adrenaline) and other catecholamines which cause muscles and fat tissue to release stored energy reserves for conversion to glucose. They also cause fat tissues to give up free fatty acids for direct oxidation. Corticoids released by the adrenals free amino acids from protein synthesis, diverting them to the liver for conversion into glucose. Another hormone secreted by the adrenals, aldosterone, raises the level of salt in the bloodstream. This causes the kidneys to produce pressors, which constrict the blood vessels in order to accommodate the faster respiration and heartbeat required for quick response. The amount of constriction controls the blood pressure; thus, the stress response is reflected in an increase in blood pressure.

Cortisone is also produced by the adrenals. It acts to suppress inflammation, which may be the result of a localized injury. The body responds to local injury with white blood cells to fight bacteria and swelling to isolate the injured area from the rest of the body and prevent foreign substances from spreading throughout the system. Thus stress places strong and conflicting demands on the immune systems; one response stimulates the body's immune system at the site of the injury (the localized response), while at the same time the immune system is suppressed away from the site (to prevent the localized inflammation from diverting all the body's resources to the injury site).

The stomach is stimulated to increase production of gastric acids. Combined with steroids, which are also produced when the generalized stress response is invoked, these acids can ulcerate the intestinal lining.

All of these responses are designed to make available enough energy to meet the stressful situation and to react to injury. The ability to react to stress is vital, but the changes which occur when the stress response is invoked disrupt some of the body's usual processes and deplete stored energy reserves. The responses in this initial stage, called the *alarm reaction*, are only the first part of what happens when the body reacts to stress.

After a demand invokes the alarm reaction, the body attempts to adapt and return to equilibrium, replenishing its defensive capacities. This is the *resistance* stage. If the body's attempts to cope with the stressor fail, or if the stressor continues beyond the body's reserve capabilities, the general adaptation mechanisms fail. This is the final stage, *exhaustion*, which leads to premature aging and the breakdown of the body. It can eventually lead to death.[14]

The stress response causes the organ systems to function at a high level, producing some wear and tear on the organs. When it is invoked frequently the body's reserves can be overtaxed, the suppression of the immune system may leave the body open to infection, and diseases such as coronary heart disease and coronary artery disease (exacerbated by the increase of blood cholesterol and blood sugar content), high blood pressure (from the release of pressors, due to increased salt content), and ulcers (caused by the combined action of released steroids and gastric acids) may result. Before such diseases develop there may be many early symptoms of chronic stress, such as loss of appetite, weight loss, irregularity, indecision and poor memory, headaches and backaches, sleeplessness, nervousness, irritability, and shakiness. At minimum, the body's response to the invocation of the stress response is fatigue.

Many types of stimuli can cause the body to react with the stress response. These stimuli are thus called *stressors*, and they may be physiological, chemical, or psychological. Light, noise, and temperature may be stressors when they are not within comfortable ranges. Physical illness, pain, and inadequate diet stress the body. Conflict and worry are examples of psychological stressors. The effects of stress are additive, so the more stressors which are present, the greater the cumulative effect. A useful model, the dynamic fatigue model, is attributed to the prominent ergonomist Etienne Grandjean (see fig. 14). According to this model, there are many daily demands on the human system. These include physical task demands, physical environment demands, intellectual demands, and psychosocial demands. The body must respond to all of these demands with coping energy in order to empty the day's accumulation of fatigue and achieve a state of recovery.

One problem with trying to use the physiological model to assess stress is the lack of a proper ruler, such as blood pressure or catecholamine level. Stress may not, in fact, be "an observable or discrete event." The experience of stress is related not to the level of the demand, but to the discrepancy between the level of the demand and the coping ability of the individual.[15] A demand which may be easily met may not provoke a stress response. The individual's perception of his or her ability (or inability) to meet the demand is crucial. If a person believes that a successful coping strategy is available, very little actual stress may result—whether the coping strategy is actually used or not. Thus the perception of control is critical in the reduction of stress. If the coping strategy is actually

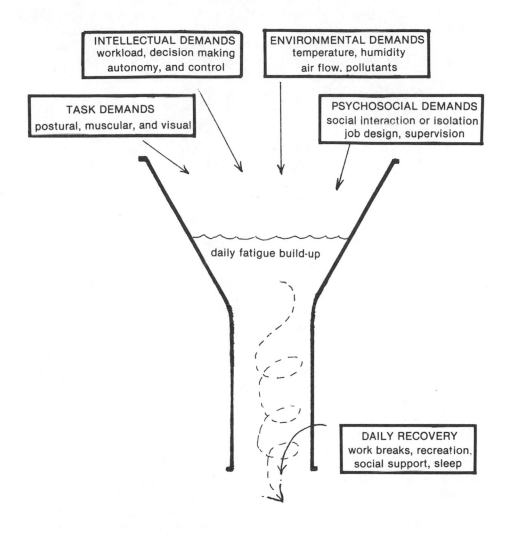

Fig. 14. Dynamic Fatigue Model (after Grandjean)

used, however, the individual must be able to tell whether or not it is working; successful coping may be as stressful as unsuccessful coping if no feedback is provided.[16]

Another way of looking at the long-term effects of stress is to assess what are called the *adaptive costs*. Coping— by which an individual either directly confronts the source of the problem or indirectly reduces the experience of stress itself — may lead to long-term problems even if the coping itself is successful. One result may be cumulative fatigue. This concept assumes that only a finite amount of coping energy is available, and that it can be depleted by prolonged use. Cumulative fatigue may be physiological, as in Selye's model, or it may take the form of psychic depletion, with such consequences as reduced attentional capacity. Long-term coping may also lead to overgeneralization, in which state the individual employs coping strategies (such as withdrawal or denial) even when not confronted by a stressor. These inappropriate responses may hamper an individual's social adjustment. If coping strategies repeatedly fail, the individual may become passive and make no attempt to cope, even in situations which would require little effort (a state known as *learned helplessness*). All of these results of coping constitute adaptive costs.

These side effects of coping may be expressed in many ways— for example, as a failure to maintain beneficial health habits. Working late to meet deadlines (coping) may disrupt nutrition and sleep habits, and excessive smoking and drinking (coping) are both harmful to health.[17]

VDT Work and Job Stress

The introduction of VDTs may expose the worker to a number of stressors. It is clear that certain types of physiological stressors which are common to office work can lead to postural and visual problems. Environmental stressors such as uncomfortable temperature ranges, inadequate ventilation, and chemical pollutants are also often present in the office environment. In addition to these physiological and chemical stressors, the introduction of VDTs may bring about other changes which can act as stressors. Stress cannot be isolated from the many other topics already discussed, because stress results from a combination of many factors, and these stressors can be additive—the effects of more than one stressor can combine to increase the individual's health risk.[18] Nor can job-related stress be separated from other stressors such as critical life events (death of a partner or divorce, for example) or from cultural and economic factors such as rates of unemployment.[19]

Some stressors which are particularly relevant to VDT work have been identified, however. The introduction of VDTs can create anxieties about job security, which may include fears that the job will change drastically or even be eliminated and concerns that the job will be de-skilled and that there will be no potential for advancement. According to Michael Smith of the Motivation and Stress section of NIOSH, "The issues of the '80s in terms of stress have to deal with job security, job de-skilling, work pressure that's brought about by technology, control over the work process and social isolation."[20]

These anxieties can be exacerbated if the workers are not included in the planning stages for implementing computerization. Computers have implications for job security, since they are usually introduced in an effort to improve productivity which, if successful, may mean that fewer employees will be needed to do the same amount of work. Fewer supervisors may be needed, too, since computers can do much of the employee monitoring that first-line supervisors previously have done. The switch from human evaluation to machine evaluation can create further anxieties in workers, who may see the computer as a kind of spy. Reinforcing this perception and significantly increasing stress among VDT operators are pay schemes that depend on computer monitoring to assess worker productivity, allowing the introduction of piece-rate pay in place of hourly pay.[21] Incentive plans, which provide bonuses for increased performance, may have much the same effect as piece-rates.

The expense of a computer system also encourages management to increase machine use by employing round-the-clock shifts, and shiftwork is known to increase worker stress. Workload, including number of hours, pacing (machine-paced or self-paced), and pressure such as deadlines also affect worker health. Machine pacing is a stressor, as is working increased hours (more than sixty hours a week may increase the risk of heart disease). System crashes can accentuate these problems since they are unpredictable interruptions which may prevent the worker from being able to complete tasks in the allotted time. "Contrary to some predictions, most employees do not regard these 'down times' as unexpected vacations. Instead, they worry that they won't be able to complete their work."[22] Where these breakdowns occur frequently, workers often respond by forcing their work pace early in the day in order to "store up" work against anticipated breakdowns. Slow system response times are also stressful and frustrating, especially if piece-rate pay schemes are in use.[23]

Workload may also be increased as more output is expected from the same number—or, more often, a smaller number—of workers. At the same time the demand for perfect output (seemingly made easy by word processing) often leads to an increased number of drafts of a document and reprocessing for minor cosmetic changes.[24]

Repetitive work, such as data entry, can produce boredom and fatigue. This is especially true if the introduction of computers leads to task fragmentation and reduces the worker's personal involvement with the completion of the final product. According to Michael Smith: "For most professional and technical jobs the computer serves as a tool that does not diminish the content of their work. However, for the majority of clerical workers, computer automation can mean more trivial tasks, more redundant activities and less use of skills and thinking processes. The effect is similar to the 'Taylorization' of blue collar jobs in the early decades of this century."[25] (Taylorism, a theory of work organization, is described more fully later in this chapter, under "Automation and the Evolution of Clerical Work.")

Social interaction may be reduced, depriving the worker of the social support which often serves as a buffer against the health effects of stress. Social support is particularly important in mitigating the effects of ulcers and neurosis. While family support is important, supervisor and co-worker support is especially important in buffering the effects of job-related stress.[26] Lack of social support at work, either from supervisors or co-workers, has been correlated to high levels of depression.[27] One of the frequently mentioned aspects of extensive VDT work is the social isolation that often accompanies it, since the operator has fewer occasions to move around and often is monitored to discourage such movement. Thus VDT operators may be denied an important support source while simultaneously being exposed to new stressors.[28]

The introduction of computers also may introduce ambiguities into job roles because the rapid nature of technological change often destabilizes old office routines. "In a number of studies, role ambiguity is the variable having the most negative (strain) consequences . . . Many employees do not know what expectations relevant people in the organization have of them as regards their performance," according to Charles J. De Wolff.[29] Computers may also reduce the scope of the worker's decision-making authority, and lack of control in one's work is a very potent stressor.[30]

Gould and Grischkowsky, who compared work with hardcopy and work with VDTs, concluded that "most complaints of fatigue and stress about computer terminals originate where workers have little or no perceived control over their work lives. . . . Ergonomics, thoughtfully designed and evaluated user interfaces, and humane management practices are required not only to arrive at exemplary workplaces but also to achieve personally fulfilling work design and organization."[31]

In one 1977 study, VDT operators had higher blood pressure levels, higher triglyceride content, and higher adrenaline secretions than non-VDT workers. These measures are all indicators of a physiological reaction to stress. A follow-up study done seven years later found that the blood pressure and triglyceride levels had essentially equalized between the two groups, an effect that the researchers attributed to the fact that during the interim many of the non-VDT users in the earlier study had added some VDT tasks and many of the VDT users in the 1977 study were actually spending less time at the VDT. This reorganization of work "demonstrates how organizational means can be used to counteract stress and health problems associated with VDT work."[32]

A survey conducted by the North Carolina Communications Workers of America reported that female VDT operators between the ages of twenty and sixty-five reported angina-type chest pain twice as often as female workers not using VDTs. Angina is a significant predictor of coronary heart disease. The survey had serious methodological flaws, including a low usable response rate (21 percent), probable reporting bias (workers with chest pain may have been more likely to participate in the study) and collaboration bias (subjects filled out surveys on their own time and may have discussed them with other workers), and a questionnaire developed for men aged thirty-five to sixty-nine which may have been inappropriate for the population being studied.[33] NIOSH reanalyzed the study data and confirmed that it indicated a link between VDT use and chest pain, but could not confirm that the chest pains reported by the workers could be clearly designated as angina.[34] These results indicate the need for further research in this area.

In spite of all these potential stressors in VDT work, it is not necessarily true that VDT work is more stressful than other types of clerical work. If task definitions, perceived control, and other known job stressors are present, it makes little difference whether the worker uses a VDT or not (although VDT users do report more eyestrain than nonusers, according to some studies).[35] Boredom, fatigue, lack of control, poor physical work environment, job ambiguity, reduced decision making, inappropriate workload, shiftwork, piece-rate work, reduced social interaction, lack of advancement potential, and job insecurity are all significant stressors,[36] but none of them is intrinsic to VDTs or computers themselves. They are instead the result of how the technology is used, and they reflect not the properties of the technology but the management style of the organization. The computer itself is neither pernicious nor benevolent; it is merely a tool for work, and its impact on the worker is determined not by technology but by the organizing principles which are used to apply the technology. "Organizational norms, management style, ergonomic factors and levels of job satisfaction seem to be more of a predictor of job stress than VDT use alone," according to Seamonds and Weiman.[37]

In fact, the introduction of the computer can have a positive impact on worker satisfaction by permitting the operator to have greater control over output and by providing immediate feedback on the display screen. This is particularly true when the task is one which has limited variety and for which the worker formerly had little or no opportunity for feedback or control. Kalimo and Leppänen investigated the work satisfaction levels of typesetters using perforators and those of typesetters using video display terminals. Perforators are input machines which provide no feedback and little opportunity to correct errors or control quality. In comparison, VDTs permit immediate feedback and allow the operator to expand the job function somewhat by correcting errors. They found that the VDT users' self-evaluations of mental activity and self-determination at work were more positive than those of the perforator typesetters.

Proofreaders and photocompositors were also investigated in this study, and similar findings correlating job satisfaction with feedback and control over work were reported in these groups. Kalimo and Leppänen conclude:

> It is obvious that computer technology makes varying and challenging work content, feedback and control possible in many ways. In too many situations the opposite unnecessarily occurs. Previously complex work is fragmented, and control over the whole work situation is diminished. Such mistakes may be a reason why many previous reports have shown negative outcomes for computerized work processes.[38]

Smith also notes that "feedback about performance is a significant aspect of worker control of the work process." He recommends increased participation by workers both in decision making and in the introduction of technological and procedural changes. These steps, combined with quality training, social interaction, clearly delineated career paths, a supportive supervisory system, and an avoidance of job downgrading or de-skilling, can be important factors in reducing VDT worker stress.[39]

Unfortunately, when VDTs are introduced, other job changes are often made which have a detrimental effect on the quality of work. The computer lends itself especially well to repetitive, routine tasks like numerical calculations, and to the atomization of tasks into subparts. It may also perform many of the mental processing functions formerly performed by the human operator, thus reducing the challenge of the job.[40] This has often encouraged the development of VDT workplaces which resemble "paper factories."[41] One survey of full-time VDT users, part-time VDT users, typists, and clerical workers who did not use office machines reported that full-time VDT users had higher job demands, more repetitive work, less autonomy, and less meaningful work than the other populations studied. The full-time VDT users also reported more stressors related to poor ergonomics, less freedom to move around, and less job satisfaction than the other groups, even though they reported that the physical nature of their offices was more satisfying.[42] Pot, Brouwers, and Padmos reported findings which seem to confirm the tendency for full-time VDT work to be more stressful. They found that autonomy decreased as time spent at the VDT increased, and that data entry provided less job autonomy than word processing.[43]

The computer has the potential to free workers from repetitive tasks like retyping and constant filing, to allow them to be more responsive to problems through rapid access to information, and to increase the mental component of clerical work by distributing the decision-making process to lower organizational levels. In some organizations the computer has already done so.[44] But it has done so only because it was introduced and implemented in ways that encouraged the use of the computer as a tool, not as a taskmaster.

Unfortunately, in far too many cases, computers are introduced into a workplace that already draws many of its assumptions about work from theories of job design that provide few opportunities for worker autonomy. In such settings the introduction of computers often exacerbates negative job components which are already present. Many of these low-challenge, low-control clerical jobs have evolved from the application of Frederick Taylor's principles of scientific management.

Automation and the Evolution of Clerical Work

Early clerical work was performed mainly by men, and it involved many of the tasks which are now thought of as managerial. Clerks had better pay, higher status, and better working conditions than factory workers, and they also had job security—clerks usually stayed at one job throughout their working lives. The growth of bureaucracies led to a much larger clerical class and the duties of clerks began to be split into two basic functions. First, the planning and supervision of the work became the basic role of a new, mostly male, managerial class; second, the routine performance of the tasks assigned by the managers became the job of a predominantly female clerical workforce.

This trend was already well underway when Frederick Taylor's theories of scientific management gained acceptance in the 1920s. His basic concept was that work can be divided into two components, the conception of tasks and the execution of tasks, and that separating conception from execution would give the greatest cost-benefit results. Higher-paid, skilled workers would not be wasted on routine work; unskilled, lower paid workers could be taught to perform simple, repetitive tasks and could be motivated to perform them efficiently by the use of pay incentives (piece-rate work).

Taylor's work was originally applied to the factory, but these principles were also easily applied to the office, where numerical data and paperwork were much easier to shift around than machinery and raw materials in a factory. Clericals were often de-skilled into pools such as typing or secretarial pools, where the entire job consisted of entering alphanumeric data on typewriters, or in low-level bookkeeping work in which tabulating machines did much of the calculating and posting. These workers were still involved in the final product of their work, even though they might do a great deal of retyping of material. With the introduction of keypunch machines, however, workers were often cut off from the final product. Their task was to enter data all day; they might never see the results. According to Fe Josefina Dy, data entry at this point had made the transition from task to occupation.[45]

These changes have taken place fairly rapidly, and the most dramatic ones— the emergence of typing, data entry, and now word processing as occupations— have happened very rapidly indeed. There is no question that the development of keypunch machines and mag card typewriters facilitated this transition. There is also no question that the nearly universal pervasiveness of the computer in the modern office can de-skill and depersonalize clerical work even more.[46] It is not surprising that many clerical workers regard the introduction of the computer with anxiety and skepticism—they have seen the impact of its technological forerunners on their pay, their status, and their emotional and physical well-being.

The de-skilling of clerical work is so common a trend in the American office that it is seldom examined separately from the technologies that made it possible. Thus the computer is frequently blamed: the computerization of work has come to mean the dehumanization of work. But in truth the introduction of the computer can provide a rare opportunity to retool office work in ways that could revitalize the workplace—when management "has the perception of the potential and the knowledge how to exploit it."[47]

In reality, however, management often does not recognize the importance of planning for the implementation of new technology. In a summary of a symposium on video displays and worker health, Brown, Dismukes, and Rinalducci note that "in most organizations, jobs are not designed—they evolve. Hundreds of thousands of dollars may be invested in a computer system with little or no attention given to its possible impact on workers and the work process."[48]

The office may be seen as a system made up of subsystems, and it is impossible to introduce change into one of these subsystems without affecting the other subsystems in some way. Goodrich defines these subsystems as follows:

> (1) the kind of people who work in the office and their psychological characteristics; (2) the work, activities, and tasks that these people perform; (3) the social processes, communications, and relationships they have with each other; (4) the organization, its structure, style, and formal and normative characteristics; (5) the type of technology that is used by people to perform their work activities; and (6) the designed environment itself.[49]

The introduction of automation will change the technology the workers use, and it will also have an effect on the physical environment. The norms of the organization will determine how automation affects people's work, as well as the skills required to perform the work. The design of the work will determine to a large extent the amount of social interaction workers will have. In light of all these interactions, organizations must plan, and plan carefully, for automation.

Reducing Stress in VDT Work

Redesigning clerical work to reduce stress and improve operator health, well-being, and productivity is no small task. There are, however, several concrete steps which organizations can take as they begin to address this issue. Operator training can improve productivity by optimizing system use and by lowering error rates, and it can simultaneously reduce stress by relieving apprehension and frustration. Thoughtfully designed software interfaces can increase operator control, provide crucial feedback, and mitigate some of the stressful aspects of system failures. The inclusion of the user in the process of selection and implementation can also increase operators' feelings of control and can provide valuable assistance in determining what kinds of features the system actually needs to have. Lastly, even the daunting task of redesigning clerical jobs can be approached in a step-by-step fashion.

Productivity and Enrichment: The Role of Training

Operating a VDT is a skilled task, whether the output is the entry of numbers in a banking system or the editorial column of the daily newspaper. As such, it requires training. Far too often, however, computers are introduced in a haphazard fashion, and no training other than the frequently inadequate manufacturer's documentation is provided. But quality training, which includes not just an orientation to the machinery but also opportunities for performance evaluation, continued learning, and personal advancement, pays off in increased productivity, reduced error rates, and reduced frustration for the operator.

Quality user training also helps to ensure that the sophisticated features of the system are used, not just the most basic ones, thus providing greater return on investment.[50] Training, combined with ergonomic improvements in software, hardware, and workstations, can increase the use of such features as middle management productivity tools by 50 percent or more, according to Franz Schneider.[51] Lack of training, however, can make workers feel uncertain, which contributes to stress.[52]

Training in the use of the system is not the only area in which training ought to be provided. VDT users should be educated about the health aspects of their jobs and about the company's specific health-related policies for users, such as rest breaks, vision correction, or pregnancy transfer. In industrial settings worker education about health and safety is considered essential, and the worker is a partner in the continued team effort to reduce accidents and absenteeism. This attitude is becoming more and more important in the office setting, particularly as operator concerns about the health and safety issues discussed in this book have gained momentum. Consultant Marilyn Joyce points out that "training in health/comfort strategies is a means of enabling workers to accept responsibility for their own well-being at work."[53]

In order to develop such an attitude of partnership, the education management provides about the issues which concern operators must be frank and thorough. Unfortunately, employer presentations to VDT workers have sometimes been perceived as shallow or conciliatory. "To be effective, education should be designed not only to inform the learner, but also to encourage critical thinking and meaningful action. However, most educational programs aimed at VDT users are corporate based and are designed more to reassure workers than to educate them."[54]

Responsive, responsible training would include education about the health effects of using VDTs, as well as training about how to control these effects. This includes instruction in such topics as the arrangement of the workstation and the work materials, vision correction and glare control, and perhaps also exercises for relieving visual and muscular fatigue.

Training should most certainly include instruction in the adjustment of chairs and workstation furniture. According to designer Niels Diffrient: "The user is going to have to have a certain amount of training to employ his or her equipment. The average office worker is a professional sitter. He or she sits all the time, and the work revolves around sitting. A professional sitter needs a professional sitting device. And a professional device always requires user training to learn how it is used properly. You don't get into an automobile and expect to drive it without training."[55] Marvin Dainoff and Leonard Mark have concluded from their field studies that operators will use furniture adjustments to achieve more comfortable postures— if they are shown how to make the adjustments.[56]

Stress reduction and stress management training may also be useful, but since these methods treat the symptoms of stress rather than the causes, their value is limited unless they are used in conjunction with efforts to reduce the causes of stress.[57]

Feedback and Control: Enhancing the Role of the User

At the very beginning, when automation is being contemplated, the users should be allowed and encouraged to participate. "The operators . . . may experience less stress if they feel that they have some degree of control over their work situation and that their needs are being considered."[58] This consultation should include operator input not only regarding workstations and physical surroundings, but also in the system which is being selected. Even though the user probably knows more about the tasks which must be performed than anyone else, users are often not included in the process until the late stages of automating the office. The result may be that a system is technically sound but is not responsive to the organization's needs.[59]

As has been mentioned above, the way that the system responds to the user can have a significant impact, positive or negative, on operator stress. Appropriate feedback from the display and the ability to correct errors are both perceived as positive characteristics, while frequent down times and long system response times have negative consequences on operator well-being. These are qualities which pertain to the *software interface*, which is essentially the bridge between the machine and the operator. An interface may be designed with attention to user needs and styles, or it may not.

Two concepts which are key to designing interfaces for users are *acceptability* and *usability*. People are adaptable, and it seems that users are more willing to adapt to the mental demands of a less-than-optimum interface than they are to poorly designed hardware; thus acceptability is a subjective concept which is difficult to evaluate. For the same reason, it is difficult to determine the costs of this adaptation by the user. But users won't bother to adapt to the system, even if it is relatively simple to use, unless it offers them usability, that is, a better way to do something they want or need to do.

Users differ as well. Some users are programmers developing systems for others; some are full-time data entry and clerical workers; some use the system only occasionally for queries. Each of these user groups will have different needs, and within these groups, users have different levels of expertise. Novice users may need the assistance provided by such features as menus, icons, or on-line help, while advanced users may find those features cumbersome and want to bypass them by using command languages. A well-designed interface ought to accommodate these different levels of expertise.[60]

For the VDT operator performing clerical tasks, generally speaking, an interface which offers the operator information about the status of a process and which allows the operator to query the system (for help, as an example) will be perceived more positively than one that does not. Slow response times and system down times are not always avoidable, but if the interface gives information to the operator ("System down for maintenance, expected up at 13:00" might be a sample message) then down time can be spent more productively, and operator stress can be reduced. Long response times may at least be acknowledged by messages such as "Working," which reassure operators that the operation to be performed has been received by the system.

Methods of entry other than the standard keyboard, such as the mouse or the light pen, may also be components of the interface. While the precise type of interface needed depends on the task to be performed, the skill of the operators performing it, and the abilities and capacities of the system on which it will be performed, the overriding principle remains the same: the interface design must give the needs of the human user a high priority. The quality of the interface should be one of the key considerations when computerization is being introduced, and actual users ought to be consulted in the process from the beginning right on through full implementation.

Redesign of the Data Entry Job

Although interface design, operator training, and good ergonomic design of the office environment are important factors in reducing operator stress, investigators have repeatedly emphasized the priority which must be given to the design of the task itself. Many of the studies cited in earlier chapters have shown that clerical workers in general experience high levels of postural and visual complaints, and surprisingly high levels of stress; in some cases, VDT operators experience these even more than nonusers, but in any case the clerical task itself seems to pose problems. The steady pattern of de-skilling clerical work by fragmenting, atomizing, and isolating it has, in too many cases, been taken one step further by the introduction of computers. Nowhere is this more obvious than in the work of the data entry operator.

Many studies have noted that certain groups of VDT users, such as editors, computer programmers, and systems developers, do not have as many health and job satisfaction complaints as data entry operators. An examination of the tasks performed in these jobs shows that many of the aspects of job design that are stressful components of other VDT tasks are not a part of these job categories. These workers, who would be categorized as professional, have more autonomy, more varied tasks, greater participation in the product, in short, a number of positive job components that are usually missing from the simple data entry task. Yet there is a great need for the data entry function to be performed, and data entry is the most prevalent type of full-time VDT work.[61] How can this occupation be redesigned or redefined to introduce or increase some of these positive components?

Certainly one option is to eliminate data entry as a full-time occupation. Since the number of health-related complaints increases with increased VDT time, there are powerful reasons to consider changing full-time data entry work to part-time work with other part-time job duties; however, this is not always possible.[62]

When data entry is to remain a full-time activity, it is important to introduce as much autonomy and control as possible into the position. This might include giving the operator responsibility for checking with document authors or clients directly when there are questions about the original source document. The tasks to be performed could be batched for completion within a reasonable time, rather than strictly scheduled by the supervisor, thus allowing the operator to schedule the work within that framework. As discussed above, the ability to get immediate feedback from the display and the freedom to correct errors and monitor quality also increase the operator's feeling of control over the work. Associating the data entry operator with the end product—either by having the operator prepare final copy or by providing verbal feedback and support (from either the supervisor or the document author)— can increase the worker's feeling of responsibility and investment in the work.

It is also important to set realistic workloads which take into consideration the difficulty of the tasks and the quality of the source documents. Frequent supportive contact from supervisors can increase social interaction in what is often a socially isolated task environment, as can direct contact with document authors. The use of computer monitoring is generally viewed as negative;[63] it

TRADITIONAL DATA ENTRY: MONOTONOUS, SOCIALLY ISOLATED, LITTLE CONTROL OVER WORK PLACE, LITTLE DECISION MAKING, VERY LITTLE OPPORTUNITY TO MOVE AROUND.

NEW DATA ENTRY: INCREASED TASK VARIETY RELIEVES MONOTONY, INCREASES PHYSICAL MOVEMENT. INVOLVING WORKER IN END PRODUCT INCREASES PERSONAL AUTONOMY, CAN INCREASE DECISION MAKING AND SOCIAL CONTACT. TEAM PRODUCTION APPROACH INSTEAD OF KEYSTROKE MONITORING INCREASES AUTONOMY AND SOCIAL SUPPORT.

Fig. 15. Redesigning the Data Entry Job

"seems to create a more competitive environment which may decrease the quality of social relationships between peers and between supervisors and subordinates."[64] Instead, a team effort approach might be useful to maximize production (see fig. 15).

For reducing stress related to muscular and visual fatigue, rest pauses are considered very important, particularly for full-time operators.

Recommendations vary, but NIOSH, for example, recommends a fifteen-minute break every two hours of VDT work with moderate visual or workload demands, and for intensive VDT work (high workload, repetitive tasks, high visual demands) fifteen minutes every hour.[65] These breaks are more effective if the operators have some discretion about when they are taken, and if they are taken before symptoms of fatigue begin.

Some union guidelines propose that VDT work be limited in duration, usually in the range of four hours per day. As noted, this may not always be possible; however, where jobs can be redesigned to provide VDT operators with other job duties (which should ideally add complexity and challenge to the overall job), it is certainly desirable to do so. This limitation on the number of hours of clerical VDT work should not, however, be used as a way to de-employ workers by making full-time jobs into part-time ones with reduced benefits, or by relying more heavily on temporary help, two trends which office workers' unions like 9to5 assert will further impoverish clerical work.

Some possibilities for added, but related, job duties for data entry operators include proofreading and editing tasks. Other nonmachine tasks might include filing and document delivery. Individual worker skills may suggest other possibilities as well. Some workers may hope to pursue a supervisory path and may be suited for some scheduling duties or team coordination tasks. Data entry positions have previously been so rigidly defined that these types of expanded duties may provide the only opportunities for advancement available to the data entry operator.[66]

In the future it is possible that new technologies will change the data entry function dramatically. Voice recognition and optical character recognition both have the potential to relieve some of the monotony of the data entry task. These technologies are not currently being used on a wide scale in clerical applications, and it will no doubt be some time before they are. In the meantime the redesign of VDT jobs for worker satisfaction and enrichment should have a high priority. If this task is attended to now, the introduction of other technologies— whatever they may be—should be less difficult, and the use of currently available technologies will be more productive for worker and employer alike.

Data entry work is only one part of the complete picture of office automation; yet if this particularly repetitive, demanding, and de-skilled task can be enriched, then certainly creative solutions are available to upgrade other dead-end jobs, whether they use VDTs or not. The task of redesigning office work is one that might be undertaken simply for the altruistic reason that it improves the quality of working life; but for more realistic reasons, it is essential to the cost-effective use of the powerful technology now available. A Hewlett-Packard team concluded that "if people are comfortable and have the proper tools to perform their jobs, there can be a significant increase in productivity, improvement in attitude and [morale] and a reduction in absenteeism and stress . . . . We are dedicated to the belief that making people happy and comfortable in their jobs is a major step toward increased productivity and significant dollar savings."[67]

Ontario Hydro's guidelines for supervisors conclude that "job output results from a balance between equipment capability, information management and human characteristics."[68] On the other hand, failure to consider the needs of the worker has a negative impact on organizational goals. According to Peter Unterweger:

Work and the social system built around it must serve as a vehicle for the attainment of some personal objectives that are not necessarily congruent with the primary purpose of production. This function of work can be ignored, but only at a price. . . . Workers will have their input into the process either explicitly through use of their intellect and by participation in decision-making or, when there is no opportunity for such, by poor performance, high absenteeism, and increasing turnover.[69]

Solutions

1. Reduce the environmental stressors and physiological stressors in the office through attention to ergonomics.

2. Bring the workers into the process so that they can participate in the introduction of automation.

3. Examine the types of tasks which will be performed and develop jobs which do not de-skill workers, but instead increase worker control and involvement. Existing jobs should also be examined for redesign.

4. Provide quality training on the use of the system, on the health and safety aspects of the work, and on ways workers can help themselves.

5. Begin the selection and/or design of the software interface early in the process, based on user needs, and include users in the evaluation and testing of the interface.

Conclusions

Stress, the body's response to the demands made on it, is a normal part of human life, but the health effects which have been linked to chronic or excessive stress can be quite serious. They include coronary heart disease, high blood pressure, ulcers, substance abuse, psychological disorders, and accidents. Physiological stress can lead to postural and visual discomfort and to more serious problems such as repetition strain injury. Medical claims, absenteeism, accidents, and poor performance due to work-related stress cost industry billions of dollars every year.

Attention to the physical workplace along with careful planning of jobs is necessary to reduce the stress of VDT work. Clerical workers in general, and therefore many VDT operators, have been shown to have surprisingly high levels of work-related stress. Much of this stress seems to be directly related to the design of the work rather than to the technology which is used to perform the work; however, computers can add to work-related stress if care is not taken when they are introduced.

VDT workers may experience anxiety about job security or job de-skilling, role ambiguity, reduced autonomy, social isolation, boredom due to repetitive work, reduced supervisory contact but increased machine supervision, increased workload, and frustration due to poor system response, poor training, and poor interface design.

None of these is intrinsic to computers themselves, but to management styles drawn from the scientific management theories of Frederick Taylor and others. Well-designed computer work, carefully introduced, can increase job satisfaction.

The most common type of VDT task, data entry, often contains many negative job design components, but even this repetitive task can be restructured so that it is enriching by increasing worker control, expanding job tasks, reducing machine pacing, and incorporating movement and rest breaks.

By improving the quality of working life companies can not only increase worker satisfaction and productivity, they can also reduce costs in absenteeism, turnover, accidents, and medical claims which are associated with the effects of stress.

Further Readings

Aronsson, Gunnar. "Work Content, Stress and Health in Computer-mediated Work: A Seven Year Follow-up Study," in *Proceedings of the International Scientific Conference: Work with Display Units*. Swedish National Board of Occupational Safety and Health. 1986. 401-404.

U.S. Department of Health and Human Services. National Institute for Occupational Safety and Health. *Job Stress Factors in Video Display Operations*, by Lawrence Schliefer and Michael J. Smith. NIOSH. 1983. Both of these papers report research results and therefore may be more technical than the casual reader would like; however, they are not difficult to understand, and they are influential works on VDT use and stress.

Cohen, Sheldon; Evans, Gary W.; Stokols, Daniel; and Krantz, David S. *Behavior, Health, and Environmental Stress*. Plenum. 1986. This comprehensive work contains chapters on the prevailing theories of stress, with particular emphasis on the adaptive costs of coping; methodological problems in field research on stress; contextual analysis of stress; and the relations between environmental stress and personal control, health, and cognitive performance. It is specialized but not overly technical and will be an excellent resource for readers who wish to study the stress concept in depth.

Dy, Fe Josefina F. *Visual Display Units: Job Content and Stress in Office Work*. International Labour Organization. 1985. This is a well-written and fascinating discussion of VDT work. Dy presents the historical context of clerical work, including a lengthy discussion of Taylorism, and then provides a detailed discussion of virtually every aspect of the data entry task. Her presentation of how data entry work can be reorganized is quite thorough and was used as a primary source for the discussion of data entry job design presented here.

Meister, Kathleen. *Health and Safety Aspects of Video Display Terminals: A Report by the American Council on Science and Health*. 2ed. American Council on Science and Health. 1985.
This slender pamphlet includes brief discussions on radiation; workstation design; glare, lighting, and vision; and standards, along with a brief but informative discussion of sources of stress in VDT work. It is designed for distribution to VDT operators and is a particularly useful example of this type of brochure. Contact the American Council on Science and Health, 47 Maple Street, Summit, NJ 07901.

Selye, Hans. "History and Present Status of the Stress Concept," in *Stress and Coping: An Anthology*, ed. Alan Monat and Richard S. Lazarus. Columbia University Press. 1985. 17-29.
Hans Selye is sometimes called "the father of stress research," and over the years his definition of stress has continued to evolve. This brief paper presents some of his most recent thinking on what stress is and what happens in our bodies when we experience stress. Other papers in this collection further expand on current thinking in the field of stress research.

6   Pregnancy and Reproduction

- Many factors, such as smoking, alcohol use, age, and prior history can increase the risk of reproductive problems.

- No direct causal link between VDT use and adverse pregnancy outcome (birth defects, spontaneous abortions, stillbirths, and so on) has been established.

- Even so, this issue still concerns many VDT operators because of reports of unusual groups or clusters of reproductive problems among operators.

- Several large-scale epidemiological studies are currently in process, and it is hoped that the results of these studies will produce more reliable results on which to base policies.

- Based on the results of earlier investigations and on early results of such studies as the Michigan epidemiological study, it seems unlikely that a direct link between VDTs and adverse pregnancy outcomes will be established.

- While experimental studies of the effects of non-ionizing radiation continue to produce inconclusive results, researchers encourage investigation of the effects of stress on human reproduction.

- Whether or not any link is established between VDTs and adverse pregnancy outcomes, operator concerns and fears may be real sources of stress and they should be regarded seriously.

Most of the complaints VDT operators report are short-term or medium-term discomforts, rather than outright injury. A few types of complaints require longer term medical intervention, but even these are generally responsive to treatment (although surgery, such as carpal tunnel surgery, is certainly something to be avoided if possible). However, some reports have alleged more serious damage. As discussed above, claims that VDT use might be implicated in the formation of cataracts have been convincingly refuted. The most serious concerns remaining, then, are the questions surrounding reports of unusual occurrences of reproductive problems among VDT operators. This chapter will investigate the various types of problems which may occur in human reproduction and some of the factors which are known to have adverse effects upon reproduction in humans. It will then present the unusual occurrences of reproductive failure which have been cited among VDT operators and discuss the investigations of those occurrences. Finally, it will describe what, to the best of current knowledge, is the apparent role of VDT work in these occurrences, if any.

Reproductive Problems Defined

Problems in pregnancy and reproduction, like other physiological problems, can range from the relatively minor to the severe. One common effect, low birth weight, correlates with several other adverse outcomes, including "congenital malformations, late fetal deaths, neonatal deaths, and sudden infant death syndrome." Low birth weight (less than 2,500 grams, or about 5 1/2 pounds) occurs in about 7 percent of livebirths.[1]

Congenital malformations (birth defects) of all types are one class of reproductive failure. These, too, can range from the disfiguring (cleft palate, clubfoot) to the life-threatening (central nervous system defects, heart defects). Birth defects occur in between 2 and 3 percent of livebirths.

Spontaneous abortions, miscarriages, stillbirths, and neonatal deaths form the most serious class of reproductive failure. There is reason to suspect that very early spontaneous abortions (before implantation of the embryo into the uterine lining, about nine days after conception) may be much more common than can be reliably investigated, since the woman may not recognize the event as an abortion, but may instead mistake it for a heavy or late menstrual period. Likewise, in self-reports of spontaneous abortions, episodes of amenorrhea (missed menstrual periods) may erroneously be reported as spontaneous abortions. Some 15 percent of recognized pregnancies are spontaneously aborted during the first twenty-eight weeks of gestation. Major genetic defects are present in at least 40 percent of all spontaneous abortions, but only 5 to 10 percent of the spontaneous abortions which occur after week twenty have such genetic defects.[2] Early spontaneous abortions in general are harder to track through medical records, since even if they are recognized they may not require medical treatment.

Between 2 and 4 percent of all full-term pregnancies result in stillbirths. All of these outcomes are, as noted above, strongly correlated to low birth weight. This class of mishap is also very interrelated, since one fetus which ultimately proves to be inviable may be carried longer, even to full term, than another fetus with similar abnormalities which might be spontaneously aborted at an early stage, depending upon a number of factors.

Another class of adverse reproductive outcome consists of predisposition to disease (childhood morbidities and malignancies, of which leukemia is one of the most common) and defects which are not detectable at birth, such as some types of mental retardation. Estimates of the incidence of severe mental retardation (about .4 percent) are based on tracking of children up to age fifteen (accurate statistics for this category are difficult to come by because of the time and expense needed to follow a population long enough to observe the effect).[3]

Multiple births with no hereditary component (due to fertility drugs or other agents), even though viable, may also be regarded as problematic pregnancy outcomes. Amniocentesis results are usually well recorded and ought also to be included in studies of reproductive incident which rely on accessible records, since elective abortions are sometimes the result of amniocentesis tests showing severe defects.

For the reasons mentioned above, as well as for reasons of study design and statistical analysis, not all occurrences of reproductive failure can be successfully investigated. Also related to the reproductive system, but not routinely included in studies of concerns about pregnancy outcome, are changes in menstrual cycles and lowered fertility.

Risk Factors Associated with Reproductive Problems

Many factors are known to increase the risk of reproductive problems, and some of these factors affect the fetus in predictable ways (see fig. 16). Cleft lip or cleft palate is most strongly associated with prior births which have the same defect; Down (or Down's) syndrome and trisomy (another type of chromosomal aberration) are both most strongly connected to maternal age. Maternal age and maternal smoking are strongly associated with spontaneous abortions, stillbirths, and lower birth weights (younger mothers are more likely to give birth to infants with low birth weights). Age is also a strong factor in neural tube (spinal cord) defects. Alcohol use is strongly associated with spontaneous abortions, as is a history of prior spontaneous abortions. Social class, sometimes in specific geographic areas, is connected to neural tube defects and is associated with lower birth weight. Ethnic or racial background has connections with certain types of problems, including such diseases as Tay-Sachs syndrome and sickle cell anemia. Maternal ill health, such as rubella, cold, or flu during the first trimester, can also have serious consequences for the fetus.

In addition, certain chemicals, drugs, and environmental agents may be directly implicated in adverse pregnancy outcome. The best known example of this is probably the drug thalidomide, which resulted in various defects, including gross deformation of the extremities. The drug DES (diethylstilbestrol) is now known to increase the likelihood of cervical cancer in female offspring, an effect which did not become apparent until these children were grown. Other drugs, including some tranquilizers and sex hormones, are also known to have adverse effects. The ability of ionizing radiation to disrupt cell growth and produce cell mutations, as well as the ability of ionizing radiation to kill cells, is established, and overexposure to ionizing radiation is well known as a causative agent in reproductive failure.

The cell damage which may result from such exposures may have many effects on the human embryo or fetus. If damage to the embryo occurs before the egg implants in the uterus, the embryo is often spontaneously aborted. Very early in embryo development, roughly the ninth day through the sixth week, the organs and limbs of the body are developed in a stage called organogenesis. Abnormal cell development at this stage may manifest itself in serious organ damage, such as congenital heart defects or structural brain damage, or in physical deformities like clubfoot or cleft palate. Abnormal cell reproduction at later stages can lead to more subtle changes, such as predisposition to disease.

FACTOR   INCREASED RISK OF

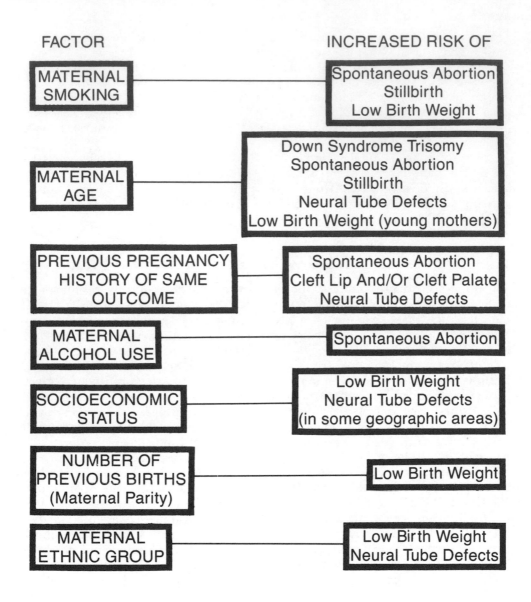

Fig. 16. Some Known Maternal Risk Factors in Human Reproduction

At this stage of development damage to the fetus may also lead to stunted growth or mental retardation.[4]

Thus a great many factors can affect pregnancy outcome—among them lifestyle, environmental exposures, quality of medical care, and heredity. In research studies these are *confounding factors*, and they must be taken into account properly or the findings will not be meaningful. Further, to be reliable, studies of reproductive failure related to paternal exposure must collect maternal histories for these factors as well.

Studying the Causes of Reproductive Problems

Usually a causal relationship between a factor, particularly an environmental agent (like radiation or thalidomide), and a disease (in this case, reproductive failure in general, or specific types of reproductive problems in particular) is discovered because groups of unusual occurrences are brought to the attention of clinicians, regulatory agencies, governments, and the like. These occurrences are investigated, and if the investigation establishes some type of pattern, further studies are performed until the evidence is clearly convincing that a particular agent does, in fact, create a risk. Research that does not yield convincing results does not prove conclusively that there is no relationship between the agent and the disease, but further research is unlikely to be undertaken when a pattern of inconclusive findings is established. The unusual occurrences which lead to these investigations are usually in the form of *clusters*, which are simply a higher than normal number of cases of disease in a small area over a limited time period.

Clusters, in themselves, do not prove that anything unusual has caused the higher than normal number of cases. The laws of probability are such that certain random groupings of events will occur by chance. When a coin is tossed, it will come up heads part of the time, and tails part of the time, and it will average out over a large number of times to about half and half; but within those many tosses, there will be unusual groups in which heads come up many times in a row, or vice versa. Clusters of events of this type are known as expected-unexpected clusters, that is, expected due to chance statistical clustering. Thus investigations must lead to findings which cannot be explained as chance occurrences in order to establish convincing relationships between agents and outcomes.

Reported Clusters of Reproductive Failure

Over the past decade, several clusters of reproductive failure have been reported among VDT operators in the United States, Canada, and Europe. Naturally, such reports have led operators to wonder if something about VDT work caused these reproductive problems, or if some other factor could be responsible. Even though these clusters have been within the expected incidence of chance groupings, many operators have found it difficult to accept that these groupings are not related to any specific cause. Because operators are concerned, numerous investigations have been undertaken to examine any possible relationship between VDT work and reproductive problems. In addition to studies of the reported clusters, epidemiological studies of other populations of VDT users have been and continue to be performed, and researchers continue to search for some agent in VDT work which could be directly related to reproductive problems. The investigations of radiation as a possible agent have been discussed above (see chap. 2). Below is a chronological summary of some of the more important cluster reports, epidemiological studies, and conclusions to date.[5]

> 1980 *Toronto, Ontario, Canada*: Reported cluster of birth defects in four out of seven pregnancies at the *Toronto Star*.
> *Marietta, GA*: Reported cluster of seven miscarriages and

three birth defects out of fifteen pregnancies between October 1979 and October 1980 (Defense Logistics Agency).
Montréal, Québec, Canada: Seven miscarriages out of thirteen pregnancies between February 1979 and February 1981 reported (Air Canada, Dorval Airport).

1981   *Atlanta, GA*: Three major birth defects and seven miscarriages reported out of fifteen pregnancies on the fifth floor of the Southern Bell offices. The NIOSH report (released in 1984) found a statistically significant increase in spontaneous abortions among women working on the fifth floor (43 percent vs. a 15 percent expected rate) but found no etiological (causative) agent. Correlation was found with job title (all adverse outcomes were non-supervisory personnel), but the relevance of this finding was unclear. Chance statistical clustering was found to be the most probable cause.[6]
Dallas, TX: Seven miscarriages and one death of a premature infant out of twelve pregnancies between May 1979 and June 1980 reported at Sears, Roebuck offices. The report from the Center for Disease Control cited chance statistical clustering as the probable cause.[7]
Toronto, Ontario, Canada: Ten miscarriages out of 19 pregnancies between 1980 and 1981 reported at the Ministry of the Attorney General.

1982   *Renton, WA*: Two birth defects and one stillbirth out of five pregnancies reported at Pacific Northwest Bell Telephone Co. between July 1980 and August 1982.
Vancouver, British Columbia, Canada: Six abnormal pregnancies out of seven between 1978 and 1982 reported at Surrey Memorial Hospital.
Ottawa, Ontario, Canada: Four miscarriages, one premature birth, and two infants born with respiratory disease reported out of eight pregnancies between 1979 and 1982 at the Office of the Solicitor General.

1984   9to5 (The National Association of Working Women) reports an increased frequency of adverse pregnancy outcomes among VDT operators responding to their 800 number VDT Hotline Survey. Survey results were criticized on grounds of selection bias, low response rate, and inadequate sample size; 6,000 self-selected callers were sent questionnaires and only 873 (15 percent) were returned.[8]
Alma, MI: Cluster of twelve miscarriages, two stillbirths, three premature births out of thirty-two pregnancies in a sixteen-month period reported at General Telephone. NIOSH to investigate.

Runcorn, England: Lee and McNamee report that 36 percent of fifty-five pregnancies at the Department of Employment end abnormally (14.5 percent miscarriages, 6.7 percent stillbirths, 22 percent malformations out of fifty-five pregnancies). The pilot study which reported these findings was based on a sample that was too small to produce reliable findings (884 female employees of the department) and was conducted primarily as an exercise in investigative techniques. Study results were skewed in part by an abnormally low rate of miscarriages in the control group (5.3 percent). The study's statistical and methodological weaknesses were so serious that the Treasury refused to publish it. Unfortunately, this refusal lead to charges of suppression and publicized the study's results far more widely than was merited.[9]

Montréal, Québec, Canada: A Canadian study (McDonald, et al.) found that heavy VDT users (20 + hours per week) have an increased frequency of miscarriage but not of birth defects or low birth weights. The overall rate for VDT users (both light and heavy) was not incompatible with what would be expected from sampling bias and principal investigator Dr. Alison McDonald attributed the observed increase to recall bias. The study was not originally conducted to investigate VDT use and adverse pregnancy outcome; it included only two questions related to VDT use.

1985 *San Francisco, CA*: NIOSH report released on the epidemiologic survey of a cluster of twenty-four problem pregnancies out of forty-eight over a five-year period at the United Airlines reservations office. The report found a 21 percent rate (which included induced abortions) which does not significantly exceed rates expected in the general population. However, the results were not considered reliable due to a response rate of only 24 percent (35 percent is considered necessary for reliable interpretation). Thus NIOSH cannot confirm the cluster as abnormal.[10]

World Health Organization working document states that emissions cannot be regarded as hazardous and that the available data do not support claims of links between VDT use and reproductive harm, although possible radiation effects cannot be completely ruled out.

Plans for a Danish epidemiological study on reproductive risks announced. Study population is 2,000 female VDT workers ages fifteen to forty who were pregnant during 1983 and 1984.

A report released by the Japanese General Council of Trade Unions found a 36 percent rate of abnormal pregnancies in its survey of VDT operators. The definition included premature births, maternal difficulties, miscarriages, and stillbirths; the broadness of the definition was thought to account for the higher rate.

Manchester, England: A cluster of six problem pregnancies out of thirteen reported at the Greater Manchester Fire Service.

University of Michigan begins a retrospective study monitoring the health and pregnancy outcome of 1,500 women employed by the State of Michigan.

Kingston, Ontario, Canada: A previously unreported cluster of six adverse pregnancy outcomes out of six (five miscarriages and one death of an infant one week after birth) at the Douglas Library at Queens University is reported by Bob DeMatteo in his book *Terminal Shock*.[11]

1986 Study underway at Mt. Sinai School of Medicine (New York), sponsored by 9to5 and the Service Employees International Union. It will examine 12,000 women over two to three years and gather data on numerous parameters. Results are expected in 1990.

The NIOSH report on the Alma, Michigan cluster shows a statistical association with VDT work but does not provide strong data that the cluster exceeds the rate in the general population, nor does it make any links between VDT work and miscarriage. NIOSH claims it will not investigate further clusters but will rely instead on the results of its proposed epidemiological study (see below).

The Canadian government disputes Dr. Hari Sharma's analysis of the problem pregnancy cluster at Surrey Memorial Hospital on the grounds that it used statistical analysis and risk assessment methods improperly and did not satisfactorily describe the protocol used for measuring emissions. Sharma's report had concluded that non-ionizing radiation was the most likely cause of the problem pregnancies.

Preliminary results from the Michigan epidemiological study show no statistically significant increase in miscarriages among VDT users. Among those who used VDTs from one to five hours per week, there were fewer miscarriages than expected; among those who used VDTs twenty-one hours or more per week, there was a slight increase in miscarriages over the expected rate. However, researchers Kelly Ann Brix and William J. Butler did not find the difference to be statistically significant, and they claim that a larger sample would be required in order to determine if this difference was due to chance. More research on 20 + hours per week users is recommended. Of the 5,739 questionnaires which were sent, 4,215 were returned. Of this group, 728 women were interviewed personally. Data were gathered on age, number of previous pregnancies, race, education, and use of alcohol and tobacco.

1987   After numerous delays and stringent budget cuts, the NIOSH epidemiological study first announced in 1982 finally gets underway. The approved study is an abbreviated, retrospective study using a population of 5,500 employees of AT&T and Southern Bell. Original plans for a retrospective/prospective study were scrapped when the control population was switched from non-VDT use to VDT use. The results are due in the fall of 1988. In spite of efforts by Office of Management and Budget (OMB) oversight subcommittee members in both the U.S. House of Representatives and the Senate to reinstate stress and fertility questions, they will not be included in the study due to budget constraints and disagreement about research methodology.

It has yet to be established that any of the clusters of adverse pregnancy outcomes represents anything more than a statistically predictable chance occurrence. Epidemiologist William Halperin of NIOSH claims that out of 100,000 groups of seventy women (which would represent a population of 7 million female VDT users), 123 of these groups would have a rate of five miscarriages over a two-year period simply by chance alone. He puts this concept even more simply: "Because VDTs are common and disease is common, clusters of disease in VDT users are inevitable."[12] The NIOSH study and other epidemiological studies may provide stronger evidence of the existence or nonexistence of unusual reproductive problems in VDT user populations.

Epidemiological Studies

Reported clusters of disease are first investigated using epidemiological studies. The purpose of an epidemiological study is to define and track particular patterns of response to some aspect of the environment. This may mean tracking outbreaks of disease or health problems, identifying the factors or organisms responsible for a previously unidentified disease (such as Legionnaire's disease), or (as is the case with VDTs) trying to determine if the reported health problems of some group are significantly different from those of the rest of the population. As is true of other types of scientific research discussed here, it is by no means clear what the results of many epidemiological studies mean. In some studies, no clear pattern has emerged; but even in those studies which claim to show statistical significance it is not wise to make the automatic assumption that the results as presented are meaningful or reliable (for a discussion of statistical significance, see chap. 2).

Epidemiological studies do not have the same goals as laboratory experiments and field studies. By definition, an epidemiological study cannot demonstrate that a specific agent caused a specific effect. It can only look at the distribution of disease in large populations of subjects over time and compare outcomes. If one group (VDT users, for example) is exposed to an agent (the VDT) and another group (non-VDT users) is not, and a particular effect is much more frequent in the first group (VDT users) than it is in the second, unexposed group—that is, frequent enough to be statistically significant—then a correlation exists between the presence of the agent (VDT use) and the occurrence of the effect. A correlation simply means that these two things—the presence of the agent and the occurrence of the effect— happened together. It does not demonstrate that that agent and only that agent caused the effect.

There are really only two results possible in epidemiological studies: 1) a statistically significant difference was observed and a correlation does exist; or 2) although no statistically significant difference was observed, the possibility of correlation between the agent and the effect cannot be ruled out. In other words, you can show that an agent and an effect are correlated if they are; but you cannot prove conclusively that they are not.

If, however, a correlation is established, properly designed experiments (laboratory experiments using animals as subjects, for example) could be done to see if a causal relationship can be demonstrated. Thus, epidemiological studies which indicate correlations generally lead to further experimental investigations. However, before attempting an experiment, the researcher should be able to suggest a plausible cause-effect sequence to investigate.

There are two basic types of epidemiological study: the *retrospective* study and the *prospective* study. In the retrospective study, the researcher attempts to go back in time to investigate what has already happened to the groups being studied (and what continues to happen during the course of the study). The prospective study begins with the present, and follows the groups being studied into the future.

The retrospective study has the advantage of allowing the investigator to inquire about events over a long time period, often years, but has the disadvantage of being less controllable than the prospective study. Because the retrospective study may rely heavily on such things as written records and self-reporting by subjects of the study, the entire picture (the environment, the suitability of matching groups over time, the changing nature of the task, for example) may not be clear. Subject self-reports may be prone to recall bias and other weaknesses common in surveys and questionnaires (see chap. 4), and available records (for example, records of spontaneous abortions kept by hospitals) may not contain all of the information needed or may require the researcher to make assumptions or extrapolations about the data that are available.

The prospective study, on the other hand, has the advantage of being (theoretically) more controllable, but has the disadvantage of requiring much more time and, therefore, expense. This is a problem especially when the question being investigated is perceived as having a high time-value; that is, a question that the affected population strongly wants to have answered immediately. The prospective study is also subject to other problems, since the groups being studied will certainly suffer from attrition (dropouts for one reason or another) and other changes (confounding factors) as the study progresses over time.

As with laboratory experiments and field studies, control group matching is important in epidemiological studies. Two types of control may be used, *cohort* and *case-control*. A cohort study matches a group that is exposed to an agent against a group that is not exposed. This grouping may be between-subjects (two separate groups) or within-subjects (one group at two different times, before exposure and after exposure). Cohort studies have one "major trap . . .

which is selecting the population not just on exposure, but because you have an inclination that they already have disease."[13] The case-control study takes matched populations, in which one group has a disease (the cases) and the other does not (the controls), and examines whether either group has been exposed to the agent being investigated.

Whichever type of control is used, it is important that the groups be large enough to show differences. The larger the study, the more likely it is that marginal effects will be detected. A study that is too small may miss even powerful effects. Thus the expected frequency of the disease in the population at large is important when determining the size of the control group; if it is a disease with low incidence, a larger control group will be necessary in order to detect the effects.

VDTs and Reproductive Risk

Epidemiological studies and cluster investigations of reproductive problems among VDT operators have been plagued with controversy over research design and methodology. Questions regarding design and methodology delayed the NIOSH study of reproductive problems and are likely to compromise the validity of its findings. Most of the investigations of clusters of adverse pregnancy outcome among VDT operators, and many of the designed studies as well, have examined populations which are too small to provide reliable results. For example, the recent Wisconsin study began with a population of nearly 6,000 operators; of these, 4,215 returned the questionnaires. But in this group of 4,215 only 728 pregnancies were reported during the time period in question. A study population of this size would not be large enough to provide reliable data about a relatively small effect.[14]

None of the inherent properties of VDTs are known to be harmful to the human fetus; at this time, none has even been clearly shown to be statistically correlated to adverse pregnancy outcomes. The results from the studies of non-ionizing radiation discussed in chapter 2 remain inconclusive, and research in this area continues to be active. However, some researchers have suggested that constrained postures due to poor workstation ergonomics might decrease blood flow to the fetus, with potentially adverse results.

Others have proposed that the type of work which is performed by many VDT operators (particularly highly repetitive, low control jobs such as intensive data entry) may lead to heightened stress levels, and that this may have an impact on reproduction. Stress has been proposed as a contributing factor in low birth weight, preterm labor, and deliveries requiring obstetric intervention, possibly because stress may affect personal habits such as sleeping, eating, and use of cigarettes, alcohol, or caffeine.[15] At least one researcher in ELF radiation, Robert O. Becker, claims that low-level ELF radiation invokes the stress response. While he does not propose this as a mechanism in adverse pregnancy outcomes, it can be inferred that such response might compound with other stressors to magnify the stress level.[16]

It has already been noted that stress effects are additive. It is difficult to demonstrate that a complex of factors, each of which individually would be insufficient to cause harm, has combined to heighten an effect. The human body during pregnancy is already in a highly stressed state. Poor work design, poor environmental and ergonomic conditions at the workplace, and the worker's own anxiety may all contribute to a stress matrix. Add to these conditions other risk factors such as smoking, age, or alcohol use, plus outside stressors such as financial difficulties or marital problems, and it is impossible to say how each factor impacts the others.

It would never occur to most people to warn a pregnant woman who smokes, who has just been fired, whose child at home has strep throat, and whose aging parent needs care that perhaps she should not drive the freeway during rush hour; but perhaps she should be warned, for this woman is subject to repeated, ongoing stress, and the drive home could literally kill.

So too the pregnant VDT operator—indeed any pregnant woman— who wants to take an active role in ensuring her own health and that of her baby should be informed about harmful agents and conditions so that she can minimize her exposure to them as much as possible. Minimizing these factors will be different for different women. Women with known physiological problems or histories of miscarriage may be advised by their doctors to leave work early in their pregnancies, no matter what their occupation. Others may be so anxious about rumors of radiation effects that they seek transfers away from VDT work in order to reduce their stress levels. Others may not need to change their work situation at all.

Solutions

1. No correlation between VDT work and adverse pregnancy outcome has been established.

2. However, VDT work should be regarded as a potential stressor. The role of stress in adverse pregnancy outcome has not been throughly researched at this time.

3. Regardless of any role that VDT work might or might not play in problem pregnancies, pregnant women who are concerned about their own health and the health of their babies can actively seek to minimize exposure to known reproductive risk factors.

4. Pending more conclusive results about the role of VDT work in reproductive failure, permitting alternate work during pregnancy (without loss of status or pay) may be the most viable way of handling operator health concerns and anxiety.

5. Such alternative work should not be mandatory, but rather should depend upon individual needs.

Conclusions

It seems likely, at this stage in the research, that other factors in the environment have a larger effect on adverse pregnancy outcome than any intrinsic property of VDTs. Even if research into the biological effects of non-ionizing radiation yields convincing evidence of a correlation between such radiation and reproductive harm, it is probable that the effect from VDTs would be considerably smaller than from other sources. In fact, researchers are beginning to suggest that any correlation between VDTs and reproductive harm is likely to have more to do with stress than with any intrinsic property of VDTs, such as radiation.

It is possible that stress may be a factor in abnormal pregnancy outcomes (possibly in combination with smoking, which is a stress-linked behavior). Therefore, stress reduction through proper job design and workplace ergonomics is a wise precaution for pregnant women and will be beneficial to workers in general as well.

Employers must remember that even though a direct causal relationship between VDT use and reproductive harm may not be shown, indirect factors such as stress level present a potential for harm which is just as real, and which must be treated seriously.

Further Readings

Brix, Kelley Ann and Butler, William J. "A Review of Proposed, Current and Completed Epidemiological Studies of VDU Workers and Pregnancy Outcomes"; Knill-Jones, R.P. "Epidemiological Approaches to Reproductive Problems in the Workplace," both in *The Second International Scientific Meeting to Examine the Allegations of Reproductive Hazards from VDUs.* Humane Technology. 1986.
Brix and Butler present a current, thorough critical review of the most important epidemiological studies to date. The authors are conducting the Michigan epidemiological study, and early results from that study are also included in a separate paper in this volume. Knill-Jones' article explains the research design and methodology of epidemiological studies. These articles are not highly technical, and together they give the reader ample background to understand and evaluate results from such studies.

VDT News: The VDT Health and Safety Report. Bimonthly newsletter.
This newletter monitors international VDT health issues, especially research related to abnormal pregnancy outcomes, very closely. Regular sections report on activity in labor, industry, and government. Articles usually provide clear information about how to obtain original source documents. It is an invaluable source for who is doing what, and what is being said.

Foster, Kenneth R. "The VDT Debate," *American Scientist* 74:163-168 (1986).
A well-balanced, accessible presentation of the two main viewpoints (labor, science and industry) on the question of radiation as a possible agent in adverse pregnancy outcomes among VDT operators.

7   Policy and Regulatory Issues

- Internationally, some VDT design and use issues are regulated, either through guidelines, standards, legislation, or union agreements.

- In the United States, no mandatory government standards for VDT design or use have been imposed. However, voluntary design standards are currently being reviewed.

- While unions, manufacturers, and employer's groups agree that better physical ergonomics can relieve many postural and visual complaints, these solutions have been introduced in only a small percentage of workplaces.

- Because VDT health and safety issues involve discomfort rather than injury, these concerns raise new questions about the definition of health at work, and about employers' responsibilities for worker well-being.

- The physical ergonomics of the workplace, the macroergonomics of the organization, and the individual behaviors of the worker must be considered together in order to develop workable answers.

It seems clear now that the controversy regarding possible radiation hazards from VDTs, in connection with either cataracts or adverse pregnancy outcomes, may very likely be a case of misplaced concern. According to Kenneth Foster, there persists a feeling that "where there's smoke there's fire" and that these concerns will be difficult to dispel: "This debate illustrates Weinberg's dictum that cognitive dissonance is all but unavoidable when the data are ambiguous and the stakes are high." Early assurances that there was no hazard did little to explain the subtleties of statistical analysis or the meaning of clusters of events, and as a result they sounded hollow and were received by workers as the vested interests' assurances that everything was just fine.[1]

There are, however, areas for genuine concern regarding health and safety, although these concerns are not as volatile as the reproduction issue. As we have seen, many visual and postural discomforts are associated with working environments which are, in many cases, inadequate for the comfortable use of

VDTs. Even more important, the design of the work itself has been haphazard, and has resulted in increased levels of stress, both physiological and psychological. These issues do not evoke the immediate attention that, for example, questions about radiation do, and the solutions are somewhat complex. It would certainly be much easier to blame VDTs and be done with it. But blaming them—or even removing them—will not alleviate the problems, most of which already existed in clerical work and have simply been accentuated by the introduction of VDTs.

Whenever a new technology is introduced, a period of adjustment is required. Just as hardware and software must be debugged, so too the human interactions involved in a pervasive new technology will soon illuminate deficiencies in many areas—physical environment, psychosocial environment, and organizational design. After two decades of increasing use of VDTs in office settings worldwide, and as many years of research, it is only now that many of these problems, potential problems, and false alarms are beginning to come into focus. Likewise, the various interested parties—labor, manufacturers, and employers—are beginning to find points of agreement and cooperation. And, as is also common when a new technology reaches a widespread level of acceptance, there is interest and concern about codifying its design and application.

Standards, Guidelines, and Legislation

There are normally four approachs to regulating the conditions of work: standards, guidelines, labor contracts, and legislation. Guidelines are often developed as an interim measure to set forth the general areas of agreement in a developing technology. Compliance is voluntary. Guidelines which prove to be usable, especially design guidelines, frequently are formalized as standards. Guidelines may also address some aspects of organizational issues, such as job design and task measurement, although these are regarded as "soft" in comparison to negotiated agreements.

Standards specify the minimum acceptable criteria which must be met to ensure reliability, safety, and compatibility. These may be industry standards which give the specifications for production (so that every manufacturer's extension cords, for example, will fit into everyone else's sockets). They may be government standards which require that certain conditions, usually design specifications, must be met if a product is to be sold to a government agency. They may be standards developed by an independent agency, such as the American National Standards Institute (ANSI). Usually, independent standards are developed by a process of draft and review in which all interested parties are invited to participate. Industry standards are usually developed by and for the manufacturers, and compliance is usually voluntary. In practice, however, products which do not meet industry standards are seldom successful in the marketplace. Government specification standards are usually mandatory, although their scope may be limited to government purchases. Independent standards are in theory voluntary but in fact usually have such impact that compliance is virtually guaranteed.

Sometimes guidelines or standards—whether industry, government, or independent—are incorporated into legislation dealing with working conditions. A number of VDT bills have been introduced in various states of the United States, but so far they have met with no success. Legislation is also often introduced based on labor contract language, and unions prepare model bills on issues which concern them. Such model bill language is more likely than standards to contain clauses about quality of working life issues related to the implementation of technology, such as rest breaks, job design, worker retraining, and the right to notification before new technology is introduced.

Standards (mandatory or voluntary) have already been adopted elsewhere, notably in Western European countries. Sweden has been particularly active in this area; since 1979 the Swedish government has had viewing standards that specify lighting, vision correction, glare control, and visual distance requirements. Swedish workers were already guaranteed prior consultation and a satisfactory working environment by the Co-determination Act of 1976, and a further 1983 National Board of Safety and Health (NBOSH) ordinance dealt with postures and working movements. A separate 1985 NBOSH ordinance dealt specifically with VDTs and addressed display characteristics, lighting, workstation design, work organization, and eye tests. It further provided for task alternation in the event that strain resulted *in spite of compliance* with all other sections of the ordinance. In the Netherlands, a 1981 amendment to the Works Council Act provides for prior consultation, worker access to experts, health and safety training, forecasting regarding staffing, and certain job design requirements.

Germany's standards organization, Deutsche Industrie Norm (DIN), developed a comprehensive standard, DIN Standard 66234, "Characteristic Values for the Adaptation of Work Stations with Fluorescent Screens to Humans." In France, VDT work is classified as one of the occupations which requires the oversight of an occupational health physician, along with consultation with workers' health, safety, and work condition committees. In Austria, night shift VDT work is subject to all the regulations which apply to "strenuous" work. In New Zealand, a Department of Labor Code of Practice (which has the force of law) requires that training, frequent short breaks, "suitable work methods," and certain eye care and ergonomic considerations be provided for those who operate VDTs more than four hours a day.[2] Australia has recently promulgated a code relating to the prevention of RSI.

As noted, the legislative standards movement in the United States has not led to any laws mandating VDT standards. In contrast to many of the Western European countries, the social and political climate in the United States is resistant to regulation of all types. In both the United States and Canada legislative efforts have been confined to the state and provincial level since national efforts at standards legislation have failed.[3] In practice, U.S. manufacturers have already retooled to comply with strong European standards such as the DIN standard. Products without a seal certifying DIN compliance cannot be sold in the Federal Republic of Germany.

The recent push for VDT legislation at the state level brings with it the possibility that many different legislative standards could be in force simultaneously, which would lead to a chaotic situation for manufacturers if substantial differences were present. This, at least, is the fear of some industry representatives, although in fact most standards bills introduced so far have been based on common language, often using union model bills.

At this time, however, none of the standards bills introduced in state legislatures in the United States has passed. The only bills to meet with success have been those which mandated informational and training measures such as the development of brochures and seminars and research reviews. Regulatory action is not always legislated, however. In 1985, Governor Toney Anaya of New

Mexico signed an executive order which sets forth strict regulations governing VDT equipment, environment, and use. It includes specific standards for selection and purchase of equipment, indoor climate, training, worker participation in the design of work areas, and rest breaks in "all governmental entities under [the Governor's] direct authority."[4]

Labor unions (notably 9to5, Service Employees International Union, and the Newspaper Guild) have been active in the standards and legislation movement because they believe that not all manufacturers and employers will voluntarily comply with guidelines or independent standards; thus, legislation requiring compliance is necessary. According to Elaine Taber, program director of 9to5, "While some enlightened employers may implement computer technology so as not to jeopardize the health and welfare of their workers, most are not going to do it without legislation."[5]

Manufacturers and employers claim that they will voluntarily comply, since it is in their best interests to do so (in order to compete in the marketplace, in the first case, and in order to increase worker productivity, in the latter). The Business and Institutional Furniture Manufacturers Association (BIFMA) standing committee on engineering standards has developed industry standards and test procedures for chairs, desks, and file cabinets, but these do not affect the installation or use of such equipment, nor do they deal with VDTs per se at all.[6]

Manufacturers and some large employers such as insurance companies are represented by the Computer and Business Equipment Manufacturer's Association (CBEMA), which opposes legislated standards. The position of retiring CBEMA president Vico Henriques concisely sums up the manufacturers' argument: "I would like to state categorically that there is no reason to consider legislation or regulation on visual displays at this time. There are no health and safety issues. There are comfort issues."[7] Other associations also oppose legislation, including the Data Processing Management Association, which calls regulatory legislation "premature and counterproductive."[8]

These groups also claim that standards would force up the cost of equipment and restrict the buyer's choice in the marketplace. They further claim that mandatory standards with specific requirements (such as workstation height dimensions) will only serve to inhibit improved ergonomic design, which could lead to worse working conditions, not better ones. According to ergonomics consultant T. J. Springer, "Once an official position is taken, the trend is away from innovation and creativity and toward conformity and compliance."[9]

An example of such a restrictive standard would be a specification that a chair must have a five-pronged base for stability. This would restrict the design of other types of chair bases which might be equally stable. Thus a standard stating that chairs must be stable at a certain load would be preferable, since any design providing such stability would be acceptable. Likewise, a specification that a 7×9 dot matrix is the minimum acceptable for character display assumes that the technology will not change rapidly. This type of standard is said to be *technology-based*. A more flexible type of standard is one which specifies what the result ought to be; for example, if a user can read the display as well as he or she can use a specified test display, then the display meets the standard, regardless of the type of presentation. This is a *performance-based* standard. Performance-based standards leave the door open for innovations employing new technologies. The subcommittee on visual information display of the International Standards Organization (an independent standards-making body) is currently developing a performance-based VDT standard.[10]

The adoption of voluntary standards in the United States may not be far off. A draft of a proposed ANSI standard has been prepared by a committee of the Human Factors Society and circulated for review. It is very likely that this document, either in its current form or in a further revision, will be adopted. This draft standard, *American National Standard for Human Factors Engineering of Visual Display Terminal Workstations*, addresses only the physical ergonomics of the hardware, support furniture, and physical environment for intensive VDT work, specifically text processing, data entry, and data inquiry. It is also expected to establish the overall direction for VDT work of all types. Issues of rest breaks, job design, pregnancy transfer, and the like, are not within its scope. Even in this limited arena, critics of the standards movement point out that guidelines currently in use—many of which have been incorporated into the proposed ANSI standard— "have little scientific reason for support."[11]

Labor Activism

European labor involvement in VDT health and safety issues has been strong for some time. The principle of co-determination has in many cases allowed workers, workers' unions, and workers' committees to participate in all aspects of VDT use including the planning and implementation stages. New technology agreements, which specify prior notification and joint consultation when new technologies are proposed for introduction, are common. In some cases, particular issues have also been legislated, but the political environment in many of these countries is such that trade union guidelines, while they do not carry the force of law, strongly affect the production and use of VDTs.

In the United States, union interest in VDT health and safety issues is comparatively recent. The perceived failure of both vendors and managers to "provide believable answers to employee questions," to strive for good design in installations or products, to consult workers about the introduction of VDTs, or to address potential hazards provided unions a potential entry point to begin organizing this rapidly expanding, relatively untapped source of members.[12] However, in spite of the controversial questions surrounding the use of VDTs, unions have not been as successful as expected in attracting members from this group. Operators of computer and peripheral equipment in the United States increased by 170 percent between 1972 and 1980,[13] and according to recent estimates only 15 percent of private sector white collar workers and 37 percent of public sector white collar workers are unionized.[14]

Typical union bargaining platforms in the United States include recommendations regarding improved lighting conditions and the use of adjustable chairs and workstations. Other issues which concern the unions include training, the right to have advance notice of the introduction of new technologies such as VDTs, assurances of job security, bans on computer monitoring, employer-paid eye examinations and vision correction, frequent rest breaks, and the right of pregnant VDT operators to transfer to non-VDT work during their pregnancy without loss of pay or seniority. Some unions would extend this to include operators who are trying to become pregnant or the spouses of women who are trying to become pregnant.

Traditional blue-collar union concerns about the introduction of automation are also receiving attention. Model contract language often includes clauses about retraining, preferential transfer, and wage increases commensurate with increased productivity. One area of concentrated union pressure is computer monitoring of work. Twenty of the more than thirty unions which are active in the VDT arena include a ban on such monitoring in their bargaining platforms. Actual contracts have generally permitted some use of monitoring, subject to certain agreements about work standards, measurement techniques, and the purposes for which the data are to be used.[15]

Union insistence on better ergonomic design of terminals was initially met with indifference by employers and manufacturers. When market forces— particularly in Europe, where labor activism on this issue had a head start— made it difficult to sell terminals with poor ergonomic design, U. S. manufacturers responded with improved designs. Employers then began to respond to union concerns about terminal design, because it was economically feasible to do so. Thus all three parties are essentially in agreement regarding some basic features of terminal design. There are also points of agreement about other aspects of the physical ergonomics of the workplace. In spite of this agreement, however, a recent nationwide study of 110 worksites found "minimum, ergonomically sound settings" for terminal use in place in only 15 percent or fewer of the sites.[16]

Even though this percentage may be expected to rise as employers recognize the benefits of providing ergonomically sound working environments, furniture alone is not the answer. According to the Westin study:

> There really are no off-the-shelf solutions to the complex ergonomic questions raised by the new technology in the workplace. *Good ergonomics is not some lucky combination of well-engineered products selected according to a magic algorithm. Rather it is the result of an ongoing process of well-informed, well-considered, experientially based management decision-making.*[17]

Part of this process of decision making involves organizational issues, such as job design, monitoring, and other quality of working life issues. These are not problems which can be solved by buying something. The solutions will require a dramatic rethinking of the way office work is structured.

Further complicating the picture is the current climate regarding occupational hazards. Scandals about the handling of workplace and nonworkplace hazards such as asbestos and toxic waste have led to a skepticism about industry's commitment to human welfare. It is largely this skepticism that has kept the controversy about radiation emissions from VDTs alive, as media treatments of industry coverups—real or fictional (*Silkwood*, *The China Syndrome*, *Ohms*)—perpetuate a belief that industry and morality are mutually exclusive and governmental coverups erode public faith. Unions, faced with declining rolls in the industrial sector, have seized on VDT health concerns as a way to unionize office workers, and have used the history of hidden hazards as an example.

Litigation

In the absence of proactive legislation, business practices are often determined by legal precedents established through litigation and workers' compensation arbitration. Some sources believe litigation involving VDTs could also become a serious problem: "Among the possible areas of litigation are product liability suits against CRT makers, malpractice suits against building architects, building engineers, and office designers."[18] In some states, such as California, it is not necessary to prove that a manufacturer was aware of any aspect of the equipment which might cause harm; it is only necessary to show that the equipment was the probable cause of the medical injury and that the equipment "did not operate in accordance with the consumer's expectations."[19]

Supervisors and employers of VDT operators are also concerned about the legal consequences of their actions. In the current climate, they often do not know the best course of action to take. Particularly sensitive is the issue of pregnancy transfer. If employers accede to worker requests, it may be perceived as an admission that VDTs can cause damage. Likewise, a special attitude toward pregnant women, and perhaps all women of childbearing age, could be seen as paternalistic or discriminatory. If such an approach were to become widespread, it could lead to serious setbacks for working women, as could a trend toward hiring part-time VDT workers who are not entitled to benefit packages.[20]

This concern appears to be justified. According to the Bureau of National Affairs, nearly all workers' compensation claims filed as of September 1984 which related to VDT use had been settled in favor of the employee.[21] However, a July 1986 report from the National Council on Compensation Insurance predicts that "pressure for rigid legislative or regulatory standards will ease in the second half of the 1980s" because all indications are that the VDT workplace is safe; no increase in worker compensation claims is expected "unless the current scientific consensus that there are no radiation-related adverse health effects is disturbed." Stress-related claims and claims related to the aging of the workforce are expected to increase generally, but VDT work itself is not expected to generate an increase in these claims.[22]

The increase in stress-related claims throughout the workforce is a relatively new trend, one which has particularly far-ranging implications for VDT work. Should employers be required to provide a workplace which minimizes stress and discomfort, or are they only responsible for keeping workers safe from injury? Should the health effects which can result from stress be defined as injuries from which workers must be protected, or are they the expected results of work, a risk which workers accept when they accept employment? How will we define health—as freedom from disease, or as a positive state of well-being? The World Health Organization explicitly describes health as "the state of complete social, psychological, and physical well-being and not the absence of disease."[23]

During the period from 1980 to 1982, claims for "work-related neuroses" doubled, while overall claims for other disabling work-related injuries went down by 10 percent. The average costs in medical and indemnity payments exceeded the average cost for other work-related diseases. Many of the working conditions experienced by VDT operators—including poor physical environment, social isolation, and lack of job control—are linked to compensable "chronic stress-related disorders."[24]

Another way of viewing this question is to examine it in terms of risk assessment. Two factors must be considered when assessing risk: 1) the probability of the consequence; and 2) the magnitude of the consequence. In the case of new technology, "the benefits of technology are no longer so irresistible that we are willing to overlook possible costs, particularly when those costs are to health and when such costs are likely to be subtle and overlooked unless assiduously examined."[25] The discomfort, stress, and behavioral effects which are found among VDT workers may originate from many causes, especially when they are likely to be the result of chronic, low-level stressors. Furthermore, the effects may appear long after the exposure and may be identical to effects which occur spontaneously or which can be the result of numerous other causes. In these circumstances it is difficult to measure how much of the problem is due to work and how much to other, nonwork stresses. Thus, "the notion of division of illness into two categories, occupational and nonoccupational, may be quite an unrealistic distinction."[26]

Productivity

If employers must protect workers from undue stress and discomfort, as well as from injury, can employers afford to make the necessary changes? Are there economic benefits for employers in providing better working conditions?

Franz Schneider found it takes only five months for the economic benefits of ergonomic furniture for a worker to pay back the initial cost. These benefits include reduced absenteeism, reduced error rates, increased machine time, reduced reports of back pain and other postural discomfort, and ultimately, increased productivity. He cites studies which report increases in productivity and performance of between 4.4 percent and 15 percent after the introduction of various ergonomic improvements.[27]

Unfortunately, industrial input-output measures of productivity simply don't translate well to the office. Office productivity may lead to such intangible—but valuable—benefits as increased overall effectiveness, improved attitudes, and renewed enthusiasm for work. It may be easier to view improvements in the quality of working life as "preventive maintenance" on a very expensive asset, the worker. For example, a twenty-year-old person who is hired at $10,000 a year, receives 8 percent raises for thirty years, receives a half-salary pension, and lives to age seventy-five represents a total investment of $2.5 million.[28]

The costs of ignoring such maintenance are as hard to evaluate as the benefits of providing it. Job-related health problems are particularly hard to compute when such problems meet the following conditions:

> They are small, but affect a large number of people;
> They are not items normally costed-out separately (e.g., input
>   errors on bills that prompt customer complaints);
> They result from physical complaints that are not serious enough
>   to trigger a trip to a physician, but are enough to make people
>   miss work;
> They are cumulative and incremental in their effects;
> The effect of occupational problems is masked by interaction with
>   other factors, such as personal health status.[29]

It is clear from previous discussions that the types of complaints reported by VDT workers fulfill every one of these conditions.

Conclusions

It is not the purpose of this work to propose that everything is a potential threat to the well-being and health of workers, or to assert that employers are expected to assume the major responsibility for employees' health. The purpose is, instead, to point out that no one thing can be isolated as the "cause" of health complaints among VDT workers. The human being is a system, made up of many subsystems, and the organizations which employ these human systems are themselves highly complex systems. While improving one thing—perhaps by providing an ergonomically designed chair—may well have positive effects, one improvement cannot be expected to provide a cure for the myriad ailments, both physical and emotional, which can afflict the worker. Thus it is important for organizations to approach automation and the changes it can bring as a systems problem.

A systems approach must also recognize that the workers' own behaviors, life stresses, attitudes, and environments affect the way they do their work and the way they feel. Individual behavior (smoking, exercise, and so on) may account for as much as 50 percent of chronic disease.[30] This is particularly important to recall when assessing the risk of adverse pregnancy outcome. Individual behaviors such as smoking, alcohol consumption, and drug use, along with individual characteristics such as age and heredity, pose far greater risks for spontaneous abortion, miscarriage, or birth defect than any of the risks which have been alleged to be due to VDT work.

At this juncture in the introduction of office automation, it is time to admit that the problems that have arisen lie not with the technology, but with the way in which the technology is used. VDTs do require certain changes in the physical workplace, and there is still room for improvement in the design of the hardware and software. But the farthest-reaching changes involve new perceptions of work design, new understandings of health and well-being, and new commitments—by employers, by workers, by unions, and by government. The interrelatedness of the problems — individual differences and lifestyles in interplay with work stresses and economic forces—requires a cooperative, rather than an adversarial, approach. Such a cooperative spirit could allow the fullest positive use of this new technology and could lead to a workplace in which employer goals for productivity and worker needs for job comfort and satisfaction could flourish together.

Further Readings

"Special Issue on Visual Display Units." *Conditions of Work Digest* 5 (1): 1986.
International Labour Office. Geneva.
Digest listings, country by country, which set forth laws and regulations, collective agreements, model codes, guidelines and policy statements pertaining to VDT use worldwide. This is an invaluable resource for anyone involved in policy formulation or labor relations regarding VDT work. Also includes an annotated bibliography on VDTs and a listing of research in progress.

U.S. Congress, Office of Technology Assessment. *Automation of America's Offices, 1985-2000.* GPO. 1985.

This document was intended to inform Congress about the likely trends in office automation for the next fifteen years, and its scope extends to include discussions of home-based work, off-shore data entry, the effects of office automation on women and minorities, and privacy and security. The chapters on office work 1985-2000, productivity and employment, and quality of worklife are particularly instructive.

Westin, Alan F., Schweder, Heather A., Baker, Michael A., and Lehman, Sheila. *The Changing Workplace: A Guide to Managing the People, Organizational and Regulatory Aspects of Office Technology.* Knowledge Industry Publications. 1985.
This comprehensive guide provides particularly good discussions of both the labor-management questions and the regulatory issues surrounding VDT use. Individual unions and their concerns are discussed in detail (including membership figures). Legal and regulatory issues are chronologically reviewed. The authors base their data on numerous resources, including their own on-site study of 110 organizations.

Notes

Notes to Chapter 1: *Introduction*

1. U.S. Congress, House, Committee on Education and Labor, Subcommittee on Health and Safety, *A Staff Report on the Oversight of OSHA with Respect to Video Display Terminals in the Workplace*, 99th Cong., 1st sess., 1985, Serial No. 99-A, Committee Print, 1. The term VDT stands for Visual Display Terminal, that is, the monitor/keyboard portion of a computer workstation. In this work, it is used to refer exclusively to cathode ray tube monitors. Although other technologies exist, cathode ray tubes are by far the most common. The term *VDU* (for Visual Display Unit) is commonly used in Europe in place of *VDT*.

2. R. Mallette, *A Supervisor's Guide to Video Display Work* (Toronto: Ontario Hydro, 1985), 18. Figures are for the United States.

3. "Numbers Worth Knowing," *VDT News* 4(3):2 (1987); Bob DeMatteo, *Terminal Shock* (Toronto: NC Press Limited, 1985), 11.

4. Grace Hopper, interview by Morley Safer, "60 Minutes" transcript 28(50), 24 August 1986, 7.

5. DeMatteo, *Terminal Shock*, 12-13.

6. Leela Damodaran, "Introduction," in *Health Hazards of VDTs?* ed. Brian Pearce (New York: John Wiley, 1984), 4.

7. Wilbert O. Galitz, "Proper Planning Stems Automation Atrophy," *National Safety and Health News*, October 1985, 96-99.

8. Stephen C. Miller, *The Eye Care Book for Computer Users* (St. Louis: Eye Care Concepts, 1986), 1; Robert T. Fertig, "VDT/Miscarriage Link Seen in Quebec Survey," *Management Information Systems Week*, 5 June 1985, 24.

9. Hans H. Tjønn, "Report of Facial Rashes among VDU Operators in Norway," in *Health Hazards of VDTs?* 17-23.

10. "NY Workers' Compensation Board Rules Cataracts Caused by VDTs," *VDT News* 2(1):7 (1985); "Workers' Compensation Update," *VDT News* 2(6):3 (1985).

11. Dave Lindorff, "Frontlines: Epidemic of Stiff Little Fingers," *Mother Jones*, May 1985, 11.

12. U.S. Congress. *A Staff Report on the Oversight of OSHA*, 12-13.

Notes to Chapter 2: *VDTs and Radiation*

1. This relationship is easy to demonstrate. Fold a thin piece of paper or a ribbon accordian-style, so that there are many folds. These folds represent waves. Mark a distance—say three inches—on a piece of paper, and compress the folded "waves" tightly, then lay the paper on the marked distance so that the edges of the waves are facing you. Count the number of folds in the marked distance. This would be the frequency. The distance between folds (measured from top to top) would be the wavelength. Now pull on both ends of the paper and observe that as the paper expands two things happen. The distance between the tops of the folds—the wavelength—gets longer; at the same time, a smaller number of folds—the frequency—will fit in the same distance. Thus the shorter the wavelength, the higher the frequency.

2. U.S. Department of Health, Education and Welfare, National Institute for Occupational Safety and Health, *Industrial Hygiene Engineering and Control, Section 6: Nonionizing and Ionizing Radiation, Student Manual* (Washington: Government Printing Office, 1978), 6.3.20-21.

3. Ibid.

4. The only studies known to this author which have shown any measurable levels of ionizing radiation are as follows:

a. A 1981 Bureau of Radiological Health study in which VDTs were tested in unshielded fault conditions in a deliberate attempt to produce the highest level of emission possible. Power supply regulation was disrupted in order to achieve maximum high voltage, a condition which would certainly not occur in actual use. In this test the units still did not emit x-radiation in excess of the standard, and in fact they failed to operate within minutes after being subjected to fault conditions. VDT emissions are covered by the television receiver emission standards; BRH tested ninety-one units between 1975 and 1980 and found that eight units emitted x-rays in excess of the standard. The three models represented by these units were either recalled for modification or were not allowed on the U.S. market. (Department of Health and Human Services, Bureau of Radiological Health, *An Evaluation of Radiation Emission from Video Display Terminals*, HHS Pub. No. FDA 81-8153 [Washington: Government Printing Office, 1981], 5-6).

b. A field study by Cox which was conducted in a location which had a high level of background radiation in several frequency ranges, including the x-ray range. These background radiation levels were erroneously reported as having been produced by the VDTs. Cox acknowledges the deficiencies of his results. (Connecticut General Assembly, General Law Committee, "VDT Radiation," a special report by the Connecticut Academy of Science and Engineering [Hartford: State of Connecticut, 1984).

5. "Lead Aprons for Operators: Help or Hindrance?" *VDT News* 1(4):16 (1984).

6. U.S. Department of Health, *Industrial Hygiene*, 6.1.8.

7. O. Bruce Dickerson and Walter E. Baker, "Health Considerations at the Information Workplace," in *Visual Display Terminals: Usability Issues and Health Concerns*, ed. John Bennett et al. (Englewood Cliffs, NJ: Prentice-Hall, 1984), 276.

8. Ibid.

9. U.S. Department of Health, *Industrial Hygiene*, 6.1.16.

10. Dickerson and Baker, "Health Considerations," 276.

11. U.S. Department of Health, *Industrial Hygiene*, 6.1.23.

12. Herbert Pollack, M.D., "Epidemiologic Data on American Personnel in the Moscow Embassy," *Bulletin of the New York Academy of Medicine* 55(11):1182-1186 (1979); U.S. Department of State, Office of Medical Services, *Evaluation of Health Status of Foreign Service and Other Employees from Selected Eastern European Posts (Final Report)* by Abraham M. Lilienfeld et al. (Washington: Government Printing Office, 1978). For an extensive discussion of this incident and its historical setting, see Nicholas H. Steneck's *The Microwave Debate* (Cambridge, MA: MIT Press, 1984).

13. In the few studies which have reported microwaves from VDTs, it is most likely that microwaves were not emitted as a fundamental frequency, but rather that microwave-range harmonics of fundamental emissions were picked up by the measuring instrument. Dickerson and Baker report microwaves at one billionth of the safety standard ("Health Considerations," 276).

14. Bruce Rupp et al., *Human Factors of Workstations with Visual Displays*, 3ed. (San Jose, CA: International Business Machines, 1984), 4; Dickerson and Baker, "Health Considerations," 276.

15. Carolyn Angell, *The Health and Safety Effects of Visual Display Units—A Review* (London: The Technical Change Centre, 1986), 11-12. The flyback transformer is usually located at the back and to one side of the terminal (see fig. 2). As a result, emissions from this source are strongest, not at the operator's position, but on the side where the tranformer is located.

16. Louis Slesin, "People Are Antennas, Too: The Biology of the Electromagnetic Spectrum," *Whole Earth Review*, Spring 1986, 51.

17. For a discussion of the development of this process, along with a discussion of the non-thermal viewpoint, the reader is referred to *The Body Electric: Electromagnetism and the Foundation of Life*, by Robert O. Becker and Gary Selden (New York: William Morrow, 1985) for Becker's account of his research in this field and the obstacles he has faced from funding agencies, colleagues, and others. Although somewhat biased, it is informative and accessible.

18. W. Ross Adey, "The Energy Around Us," *The Sciences* 26(1):54 (1986).

19. The differences in acceptable levels between East and West may be due partly to the different methods used to measure the fields and partly to differences in scientific approach. For instance, Eastern bloc scientists have been interested not only in readily observable effects such as birth defects, but also in more subtle effects such as behavioral

changes. For a discussion of these measurements and other differences, see D.H. Sliney, M.L. Wolbarsht, and A.M. Muc, "Editorial: Differing Radiofrequency Standards in the Microwave Region—Implications for Future Research," *Health Physics* 49(5):677-683 (1985).

20. Barbara S. Brown, Key Dismukes, and Edward J. Rinalducci, "Video Display Terminals and Vision of Workers: Summary and Overview of a Symposium," *Behaviour and Information Technology* 1(2):121-140 (1982).

21. A. Çakir, D.J. Hart, and T.F.M. Stewart, *Visual Display Terminals: A Manual Covering Ergonomics, Workplace Design, Health and Safety, Task Organization* (New York: John Wiley & Sons, 1980), 218.

22. "NY Workers' Compensation Board Rules," 7-8; "Workers' Compensation Update," 3.

23. "Arbitrator Finds No Health Hazard in Use of Video Display Terminals," *American Newspaper Publishers Association, Labor and Personnel Relations Bulletin*, no. 6496:77-83 (March 3, 1978).

24. National Research Council, Committee on Vision, *Video Displays, Work, and Vision* (Washington: National Academy Press, 1983), 58-60.

25. International Commission on Illumination, *Vision and the Visual Display Unit Workstation* (Paris: CIE, 1984), 8.

26. Milton M. Zaret, "Cataracts and Visual Display Units," in *Health Hazards of VDTs?* 47-60.

27. Helen Feeley, *The VDT Operator's Problem Solver* (Ottawa, Ontario: Planetary Association for Clean Energy, 1984), 8-9.

28. A. Çakir, "Are VDTs a Health Hazard?" (Paper presented at the World Conference on Ergonomics in Computer Systems, Los Angeles, September 1985), 7.

29. National Research Council, *Video Displays, Work, and Vision*, 3.

30. Recalling the earlier demonstration of wavelength and frequency using folded paper, it can be seen that by pushing and pulling on the paper the shape of the wave can be changed; for example, by elongating one side of the paper wave and pushing in on the other, a slow rise and sharp fall can be produced; this sawtooth shape is roughly characteristic of some of the VDT emissions researchers are interested in.

31. Robert Arndt and Larry Chapman, *Potential Office Hazards and Controls* (Washington: Office of Technology Assessment, 1984), 100-101.

32. "Non-Ionizing Radiation Biohazards and VDTs," *Office Systems Ergonomics Report* 3(5):9 (1984).

33. "VDT Radiation: New Research Suggests Biological Effects Possible," *VDT News* 1(5):6 (1984).

34. "ONR PEMF Project Delayed," *Microwave News: A Report on Non-Ionizing Radiation* 6(3):11 (1986). These experiments will be performed by: Jocelyn Leal (of the original Delgado team in Spain); Alexander Martin (U. of Western Ontario); Kjell Hansson Mild (National Board of Occupational Safety and Health [NBOSH], Sweden, one of the principal researchers on the controversial Swedish mice study); William Koch and Ezra Berman (U. of California and EPA), and Arnold Fowler and Jack Monahan (National Center for Device and Radiological Health, FDA). "EPA Ready to Ship Equipment to Labs for Delgado-Like Tests," *COT Center Report* 2(16):2 (1986).

35. Bernhard Tribukait, Eva Cekan and Lars-Erik Paulsson, "Effects of Pulsed Magnetic Fields on Embryonic Development in Mice," in *Work with Display Units 86*, ed. Bengt Knave and Per-Gunnar Widebäck (Amsterdam: North-Holland, 1987), 131.

36. "Swedish Mice Study: Effects Still Significant," *Microwave News: A Report on Non-Ionizing Radiation* 6(3):10 (1986).

37. "VDT Radiation Again Linked to Abnormal Pregnancies in Mice," *VDT News* 4(4):9 (1987).

38. H. Mikołajczyk et al., "Effects of TV Sets Electromagnetic Fields on Rats," in *Work with Display Units 86*, 122-128.

39. H. Mikołajczyk et al., "Task-Load and Endocrinological Risk for Pregnancy in Women VDU Operators," in *Work with Display Units 86*, 115-122.

40. "PEMFs and Chick Eggs: Results Mixed, Mechanism Elusive," *Microwave News: A Report on Non-Ionizing Radiation* 6(3):1 (1986).

41. Robert O. Becker and Andrew A. Marino, *Electromagnetism and Life* (Albany: State University of New York Press, 1982), 188; T. Terrana, F. Merluzzi, and E. Guidici, "Electromagnetic Radiations Emitted by Visual Display Units," in *Ergonomic Aspects of Visual Display Terminals*, ed. E. Grandjean and E. Vigliani (London: Taylor & Francis, 1980), 16-17.

42. Sliney, Wolbarsht, and Muc, "Editorial: Differing Radiofrequency Standards" (see chap. 2, n. 19), 679.

43. K.S. Shroder-Frechette, *Risk Analysis and Scientific Method* (Boston: D. Reidel, 1985), 3.

44. World Health Organization, *Radiofrequency and Microwaves: Environmental Health Criteria 16* (Geneva: WHO, 1981), 83.

45. Sliney, Wolbarsht, and Muc, "Editorial: Differing Radiofrequency Standards," 678.

46. William E. Murray, "Is There A Radiation Risk?" in American Industrial Hygiene Association, *Health and Ergonomic Considerations of Visual Display Units — Symposium Proceedings* (Akron, OH: American Industrial Hygiene Association, 1983), 89-90.

47. Adey, "The Energy Around Us," 54.

48. "Viewpoint," *At the Centre* 9(5):21-22 (1986).

49. Eileen Senn Tarlau, "Measurement and Shielding of VLF Radiation from Video Display Terminals," *Journal of the American Industrial Hygiene Association* 47(1):A8-A10 (1986).

50. Arndt and Chapman, *Potential Office Hazards and Controls*, 102, 105.

Notes to Chapter 3: *Eyes and Vision*

1. Lawrence Rose, MD, "Workplace Video Display Terminals and Visual Fatigue," *Journal of Occupational Medicine* 29(4):321 (1987).

2. C.N. Ong, "VDT Work Place Design and Physical Fatigue: A Case Study in Singapore," in *Ergonomics and Health in Modern Offices*, ed. Etienne Grandjean (London: Taylor & Francis, 1984), 486-487.

3. F.W. Campbell and K. Durden, "The Visual Display Terminal Issue: A Consideration of Its Physiological, Psychological and Clinical Background," *Ophthalmic and Physiological Optics* 3(2):175 (1983).

4. Campbell and Durden, "The Visual Display Terminal Issue," 190; John D. Gould and Nancy Grischkowsky, "Doing the Same Work with Hard Copy and with Cathode Ray Tube (CRT) Computer Terminals," *Human Factors* 26(3):323-337 (1984); Karl Gösta Nyman, "Refraction in VDU Operators — A Comparison with Other Professions," in *Work with Display Units 86*, 556-561.

5. Steven J. Starr, "Effects of Video Display Terminals in a Business Office," *Human Factors* 26(3):356 (1984).

6. National Research Council, *Video Displays, Work, and Vision*, 143.

7. Peter Alan Howarth and Howell Owen Istance, "The Association between Visual Discomfort and the Use of Visual Display Units," *Behaviour and Information Technology* 4(2):131-149 (1985).

8. Finn Levy and Ingrid Greger Ramberg, "Eye Fatigue among VDU Users and Non-VDU Users," in *Work with Display Units 86*, 42-52.

9. W. Bachmen et al., "Occupational Health Aspects of Work with Display Units," in *Proceedings of the International Scientific Conference: Work with Display Units* (Stockholm: Swedish National Board of Occupational Safety and Health Research Department, 1986), 983.

10. Cakir, Hart, and Stewart, *Visual Display Terminals*, 215.

11. Leif R. Hedman and Valdimar Bhiem, "Short-Term Changes in Eyestrain of VDU Users as a Function of Age," *Human Factors* 26(3):357-370 (1984).

12. Steven Sauter, L. John Chapman, and Sheri J. Knutson, *Improving VDT Work: Causes and Control of Health Concerns in VDT Use* (Lawrence, KS: The Report Store, 1986), 25, 42-44.

13. Humidity and static are discussed in greater detail in chapter 4.

14. Campbell and Durden, "The Visual Display Terminal Issue," 186.

15. Feeley, *The VDT Operator's Problem Solver*, 15.

16. Howarth and Istance, "Association between Visual Discomfort," 136; Tsuneto Iwasaki and Shinji Kurimoto, "Objective Evaluation of Eye Strain using Measurements of Accommodative Oscillation," *Ergonomics* 30(3):581-587 (1987).

17. Jan de Groot and Andre Kamphuis, "Eyestrain in VDU Users: Physical Correlates and Long-Term Effects," *Human Factors* 25(4):409-415 (1983).

18. Peter Alan Howarth and Howell Owen Istance, "The Validity of Subjective Reports of Visual Discomfort," *Human Factors* 28(3):347 (1986).

19. Çakir, "Are VDTs a Health Hazard?" 14.

20. Recommendations in this section have been drawn from several sources: the Human Factors Society, *American National Standard for Human Factors Engineering of Visual Display Terminal Workstations Revised Review Draft* (Santa Monica, CA: Human Factors Society, July 1986); Trades Union Congress, *TUC Guidelines on VDUs* (London: Trades Union Congress, 1985); Çakir, Hart, and Stewart, *Visual Display Terminals*; and Marvin J. Dainoff and Marilyn Hecht Dainoff, *People & Productivity: A Manager's Guide to Ergonomics in the Electronic Office* (Toronto: Holt, Rinehart & Winston of Canada, 1986).

21. Çakir, "Are VDTs a Health Hazard?" 12.

22. The terms usually used to describe polarity, such as negative contrast, positive contrast, positive polarity, and negative polarity, are easy to confuse; I have chosen to adopt the phrase *reverse video* for darker character/lighter background displays. I use the descriptive phrase *lighter character/darker background* to refer to the display polarity which is most commonly used.

23. Hans-Henrik Bjørset, "Lighting for Visual Display Unit Workplaces," in *Work With Display Units 86*, 331-339; Roger I. Wibom and Lars W. Carlsson, "Work at Visual Display Terminals among Office Employees: Visual Ergonomics and Lighting," in *Work with Display Units 86*, 357-367.

24. Lawrence M. Schleifer and Steven L. Sauter, "Controlling Glare Problems in the VDT Work Environment," *Library Hi Tech* 3(4):21 (1985).

25. Çakir, "Are VDTs a Health Hazard?" 15.

26. Schleifer and Sauter, "Controlling Glare Problems," 21.

27. Brown, Dismukes, and Rinalducci, "Video Display Terminals and Vision of Workers," 129.

28. Schleifer and Sauter, "Controlling Glare Problems," 21-25.

29. P.R. Boyce, *Human Factors in Lighting* (New York: MacMillan, 1981), 289-294.

30. Hardwin Tibbs, *The Future of Light* (London: Watkins, 1981), 83-87.

31. Isaac Turiel, *Indoor Air Quality and Human Health* (Stanford, CA: Stanford University Press, 1985), 119.

32. George Rand, "Caution: The Office Environment May Be Hazardous to Your Health," *AIA Journal* 68(12):41 (1979).

33. Tibbs, *The Future of Light*, 51-70.

34. Sauter, Chapman, and Knutson, *Improving VDT Work*, 34-37.

35. Optometrists receive extensive training in refraction and are best qualified to prescribe eyeglass corrections. Ophthalmologists are physicians who specialize in diseases of the eye. In most cases, an optometrist can refer patients to an ophthalmologist if he or she finds any abnormalities that suggest disease or injury. However, in some states optometrists are not allowed to perform certain diagnostic tests, so it is best to check this out with the optometrist or physician. Opticians, like pharmacists, are technicians who can fill prescriptions but cannot make diagnoses or prescribe corrections.

36. Rose, "Workplace Video Display Terminals and Visual Fatigue," 323.

37. Somewhere between one in 2,500 and one in 10,000 people are affected by photosensitive epilepsy, a form of the disease in which seizures are triggered by certain types of visual stimuli, notably lights which flicker or flash at certain frequencies. It is possible that VDT viewing could trigger seizures in these individuals, but it is not likely

that VDT viewing at work would produce a first-time seizure. Epilepsy usually manifests itself during adolescence, and most photosensitive epileptics will already be aware of the condition by the time they enter the workforce. In addition, frequency and pattern are important in triggering seizures, and many VDTs would not present the right conditions. A few variables, such as flicker, jitter, raster line visibility, and the lower frequency refresh rate used when interlacing is employed, might make some VDTs more likely to trigger seizures than others. However, since many photosensitive epileptics respond well to anti-convulsive medication, even these individuals may be able to use VDTs safely. Each individual is different, hence the strong recommendation for a professional consultation. (Çakir, Hart, and Stewart, *Visual Display Terminals*, 219-220; "Ramazzini's Corner: VDTs and Epilepsy," *At the Centre* 9(6):17 [1987]).

38. House Committee on Education and Labor, *A Staff Report on the Oversight of OSHA*, 18.

39. Barry J. Barresi, O.D. and Jesse Rosenthal, O.D., *New York State Occupational Vision Benefit Plan Study: An Evaluation of a Vision Plan for VDT and Office Workers* (New York: State University of New York, Center for Vision Care Policy, 1986), 19-20.

Notes to Chapter 4: *Strain and Injury*

1. *Office Systems Ergonomics Report* 5(5):16 (1986).

2. Steven L. Sauter et al., "Chronic Neck-Shoulder Discomfort in VDT Use: Prevalence and Medical Observations," in *Proceedings of the International Scientific Conference*, 154.

3. Çakir, "Are VDTs a Health Hazard?" 3.

4. Sauter, Chapman, and Knutson, *Improving VDT Work*, 11.

5. Ibid., iii.

6. Thomas J. Armstrong et al., "Repetitive Trauma Disorders: Job Evaluation and Design," *Human Factors* 28(3):325-336 (1986).

7. Gabriele Bammer and Ilse Blignault, "A Review of Research on RSI," in Australian National University, *ANU Research on RSI: Proceedings*, 5.

8. Robert Arndt, "Body Posture," in *Health and Ergonomic Considerations of Visual Display Units*, 31.

9. Bammer and Blignault, "A Review of Research," 5-9.

10. Radford C. Tanzer, "The Carpal Tunnel Syndrome: A Clinical and Anatomical Study," *The Journal of Bone and Joint Surgery* 41A:626-634 (1959).

11. Sauter, Chapman, and Knutson, *Improving VDT Work*, 28, 32.

12. Thomas Läubli, "Review on Working Conditions and Postural Discomfort in VDT Work," in *Proceedings of the International Scientific Conference*, 3-6.

13. Choon-Nam Ong and Wai-on Phoon, "Influence of Age on Performance and Health of VDU Workers," in *Work with Display Units 86*, 214.

14. Rachel Blau, "Repetitive Strain Injury: A New High Tech Epidemic,".*Video Views* 3(2):1 (1986).

15. Gabriele Bammer, "Musculo-Skeletal Problems Associated with VDU Use at the Australian National University: A Case Study of Changes in Work Practices," in *Trends in Ergonomics/Human Factors III: Proceedings of the Annual International Industrial Ergonomics and Safety Conference*, ed. Waldemar Karwowski (Amsterdam: North-Holland, 1986), 285-293.

16. Tony Horwitz and Geraldine Brookes, "Chronicle: 'Buggered for Life'—by VDTs," *Columbia Journalism Review* 24(2):10-12 (1985).

17. Susan Rowe, Maurice Oxenburgh, and David Douglas, "Repetition Strain Injury in Australian VDU Users," in *Work with Display Units 86*, 40.

18. "Concern Grows over RSIs," *VDT News* 4(1):3 (1987).

19. Bill Davis, "Carpal Tunnel Syndrome Strikes VDT Operators," *Video Views* 2(3):3 (1985).

20. Diana Hembree and Sarah Henry, "A Newsroom Hazard Called RSI," *Columbia Journalism Review* 25:19-24 (1987).

21. Robert Arndt, "Working Posture and Musculoskeletal Problems of Video Display Terminal Operators—Review and Reappraisal," *Journal of the American Industrial Hygiene Association* 44(6):437 (1983).

22. Ibid., 437-446.

23. Martin G. Helander, Patricia A. Billingsley, and Jayne M. Schurick, "An Evaluation of Human Factors Research on Visual Display Terminals in the Workplace," in *Human Factors Review 1984*, ed. Frederick A. Muckler (Santa Monica, CA: Human Factors Society, 1984), 55-129.

24. Institute de recherche en santé et en sécurité du travail du Québec, *Report of the Task Force on Video Display Terminals and Workers' Health* (Montréal: IRSST, 1985); National Research Council, *Video Displays, Work, and Vision*; Çakir, Hart, and Stewart, *Visual Display Terminals*.

25. It has been argued that the introduction of a questionnaire itself is an intervention, and thus constitutes a change in the environment which may confound results.

26. There are many good discussions of the Hawthorne effect, so named because of the Western Electric plant in Hawthorne, Illinois, where the effect was first recorded. One such source is H.A. Landsberger, *Hawthorne Revisited: Management and the Worker, Its Critics and Developments in Human Relations in Industry* (Ithaca, NY: Cornell University Press, 1958).

27. National Research Council, *Video Displays, Work, and Vision*, 134.

28. Etienne Grandjean, *Ergonomics in Computerized Offices* (London: Taylor & Francis, 1987), 119-120.

29. Scott Kariya, "Seeking a Perfect Chair," *PC Magazine*, 2 October 1984, 140.

30. E. Grandjean, W. Hünting, and M. Pidermann, "VDT Workstation Design: Preferred Settings and Their Effects," *Human Factors* 25(2):161-175 (1983); Läubli, "Review on Working Conditions," 3-6.

31. A.C. Mandal, *The Seated Man: Homo Sedens* (Denmark: Dafnia, 1985).

32. Marvin J. Dainoff and Leonard S. Mark, "Task and the Adjustment of Ergonomic Chairs," in *Work with Display Units 86*, 294-302.

33. Recommendations in this section have been drawn from the following sources: Sauter, Chapman, and Knutson, *Improving VDT Work*; National Research Council, *Video Displays, Work, and Vision*; Trades Union Congress, *TUC Guidelines on VDUs*; Human Factors Society, *American National Standard*; Çakir, Hart, and Stewart, *Visual Display Terminals*; Dainoff and Dainoff, *People & Productivity*.

34. Human Factors Society, *American National Standard*, 8.7.6.

35. Läubli, "Review on Working Conditions," 3-6.

36. Tom Stewart, "VDU Ergonomics—A Review of the Last Two Years," in *Health Hazards of VDTs?* 43.

37. Dainoff and Dainoff, *People & Productivity*, 38-42.

38. Grandjean, Hünting, and Pidermann, "VDT Workstation Design," 161-175.

39. Human Factors Society, *American National Standard*, 8.4-8.6.

40. Armstrong, "Repetitive Trauma Disorders," 333.

41. Sauter, Chapman, and Knutson, *Improving VDT Work*, 30.

42. Steven L. Sauter et al., "Wrist Trauma in VDT Keyboard Use: Evidence, Mechanisms and Implications for Keyboard and Wristrest Design," in *Proceedings of the International Scientific Conference*, 230-233.

43. "The Ergonomics of Office Keyboards," *Office Systems Ergonomics Report* 5(2):7-22 (1986).

44. Jan-Erik Hansson and Monika A. Attebrant, "The Effect of Table Height and Table Top Angle on Head Position and Reading Distance," in *Proceedings of the International Scientific Conference*, 422.

45. Jennifer Evans, "Questionnaire Survey of British VDU Operators," in *Proceedings of the International Scientific Conference*, 565-569. VDTs are generally not the source of excessive noise. In fact, they are quieter than the typewriters they often displace. The noise from transformers or cooling fans in VDTs is generally not annoying except to some individuals who are very sensitive to high frequencies, or those performing intense concentration tasks. These ultrasound emissions are thought to be

safe, but children (whose hearing is generally more sensitive than adults) and adults with excellent hearing may find these sounds irritating. However, devices associated with VDTs such as printers are more immediate sources of disturbing noise levels, and they should be dampened either by being isolated or by the use of acoustic covers. The overall levels should be below 65 dB, and below 55 dB for intense concentration task areas, according to Çakir, Hart, and Stewart (*Visual Display Terminals*, 187). In many offices which use an open plan system in which partitions of various heights are used instead of walls, noise can be a major factor in contributing to stress, feelings of loss of privacy, and feelings of lack of control. VDTs, because they are quieter than most other office devices, may actually make the lack of privacy more noticeable, since conversations are no longer masked by competing noise. In some cases white noise (neutral noise with no information content) has been introduced to provide a masking effect.

46. Çakir, Hart, and Stewart, *Visual Display Terminals*, 221.

47. Michael J. Suess, "Health Impact of Work with Visual Display Terminals," in *Work with Display Units 86*, 7-8; Çakir, Hart, and Stewart, *Visual Display Terminals*, 181. Newer models, however, are generally in the lower end of the range in terms of heat production.

48. Campbell and Durden, "The Visual Display Terminal Issue," 175-192. *Relative humidity* describes the percentage of moisture compared to a saturated atmosphere. A relative humidity of 100 percent means that the air contains the maximum amount of moisture it can hold at a specified temperature.

49. Çakir, Hart, and Stewart, *Visual Display Terminals*, 222.

50. John D. Spengler and Ken Sexton, "Indoor Air Pollution: A Public Health Perspective," *Science* 221(4605):9-17 (1983).

51. Trades Union Congress, *TUC Guidelines on VDUs*, 17.

52. "We believe that the facial rash which has been seen in Norway is in fact an occupational contact dermatitis, caused by irritating submicron dust particles precipitating on the skin of the VDU operator if he is accumulating static electricity." Hans H. Tjønn, "Report of Facial Rashes," 17-23.

53. European Computer Manufacturers Association, *Visual Displays: Health Aspects* (Geneva: December 1985), 18.

54. Walter Cato Olsen, "Facial Particle Exposure in the VDU Environment: The Role of Static Electricity," in *Proceedings of the International Scientific Conference*, 797-800.

55. Carola Lidén and Jan E. Wahlberg, "Does Visual Display Terminal Work Provoke Rosacea?" *Contact Dermatitis*, 13:235-241 (1985); Victor Lindén and Sturla Rolfsen, "Video Computer Terminals and Occupational Dermatitis," (letter to the editor), *Scandinavian Journal of Work, Environment & Health* 7:62-64 (1981).

56. Olsen, "Facial Particle Exposure," 799.

57. Ulf OV Bergqvist, "Video Display Terminals and Health: A Technical and Medical Appraisal of the State of the Art," *Scandinavian Journal of Work, Environment & Health* 10(Supplement 2):57-59 (1984).

Notes to Chapter 5: *Stress*

1. Hans Selye, "History and Present Status of the Stress Concept," in *Stress and Coping: An Anthology*, ed. Alan Monat and Richard S. Lazarus (New York: Columbia University Press, 1985), 17-29. Selye's definition states that these demands are nonspecific, but many alternate concepts of stress state that specific demands may elicit specific responses.

2. U.S. Congress, Office of Technology Assessment, *Automation of America's Offices, 1985-2000* (Washington: Government Printing Office, 1985), 20.

3. R.J. Tinning and W.B. Spry, "The Extent and Significance of Stress Symptoms in Industry—with Examples from the Steel Industry," in *Stress, Work Design, and Productivity,"* 129-130.

4. "Major Unions Mapping Strategy on Occupational Stress Issue," *The Office Health & Safety Monitor* 1(1):1 (1986).

5. "Current Trends: Leading Work-Related Disease and Injuries," *Morbidity & Mortality Weekly Report* 35(39):619 (1986).

6. Norman Bodek, quoted in "DEMA Survey Cites Problems among Data Entry Operators," *The Office Health & Safety Monitor* 1(1):3 (1986).

7. "Major Unions Mapping Strategy," 2.

8. "The question of a possible reproductive hazard associated with video screen work is by no means finally settled. We think that our data show that no marked risk is likely to be involved, especially not for spontaneous abortion, while the possibility of an effect on the rate of birth defects cannot be ruled out. If present, the increased risk is most likely an effect of co-varying risk factors like stress and smoking, which at least increase the risk of premature birth with low birth-weight and of perinatal mortality." Anders Ericson and Bengt Källen, "An Epidemiological Study of Work with Video Screens and Pregnancy Outcome: II. A Case-control Study," *American Journal of Industrial Medicine* 9(5):473 (1986).

9. Michael J. Smith, Pascale Carayon, and Kathleen Miezio, "VDT Technology: Psychosocial and Stress Concerns," in *Work with Display Units 86*, 695-712.

10. J.A. Bonnell, "VDTs and Health: Fact or Fancy?" in *Work with Display Units 86*, 3-5; Ricardo Edström, "Regulating VDU Work Places," in *Proceedings of the International Scientific Conference*, 987.

11. Çakir, "Are VDTs a Health Hazard?" 18.

12. The increased cost of energy, particularly since the energy crisis of 1973, has led to the development of an energy- conserving building design, the so-called "tight building." The reduced ventilation which characterizes these buildings can permit the buildup of contaminants. Substances commonly used in the construction and operation of the modern office have been shown to have health effects, such as the formaldehyde used in insulation and building materials, chemicals used in carbonless copy paper (CCP), and ozone produced by photocopiers. Poor ventilation, higher than normal temperatures, and low relative humidity may aggravate the effects of these irritants. Sometimes pollutants or other chemical or biological agents may produce discomfort throughout an office building. Headache, eye irritation, shortness of breath, chest tightness, dizziness, nausea, fatigue, and throat irritation are common symptoms of "building sickness." Some building outbreaks are bacterial or microbial in origin. The outbreak of Legionnaire's disease in a Philadelphia hotel in 1976 was traced to a bacterium which spread through the air vents. Maintenance of the HVAC (heating, ventilation, and air conditioning) system is important in the prevention of "sick buildings." Often, increasing the ventilation rate of a building will produce a dramatic reduction in symptoms caused by airborne contaminants. In cases involving actual building infection, sterilization, replacement of affected units (humidifiers, air conditioners, and the like), and increased ventilation will usually show improvement.

13. A good review of the most commonly held theories of stress is presented in Sheldon Cohen et al., *Behavior, Health, and Environmental Stress* (New York: Plenum, 1986), 1-23. Another valuable review of stress models is Tom Cox and Colin Mackay, "A Transactional Approach to Occupational Stress," in *Stress, Work Design, and Productivity*, 91-113.

14. Selye, "History and Present Status," 17-29.

15. Tom Cox, "The Nature and Measurement of Stress," *Ergonomics* 28(8):1156 (1985).

16. Cohen, *Behavior, Health, and Environmental Stress*, 1-23.

17. Ibid., 8-10.

18. Fe Josefina F. Dy, *Visual Display Units: Job Content and Stress in Office Work* (Geneva: International Labour Office, 1985), 63-64.

19. Charles J. De Wolf, "Stress and Strain in the Work Environment: Does It Lead to Illness?" in *Behavioral Medicine: Work, Stress and Health*, ed. W. Doyle Gentry, Herbert Benson, and Charles J. De Wolff (Norwell, MA: Martinus Nijhoff, 1985), 33-43; "Current Trends," *Morbidity & Mortality Weekly*, 619; Suess, "Health Impact of Work with Visual Display Terminals," 6-15.

20. "Major Unions Mapping Strategy," 2.

21. U.S. Department of Health and Human Services, National Institute for

Occupational Safety and Health, *Job Stress Factors in Video Display Operations*, by Lawrence Schleifer and Michael Smith (Cincinnati, OH: NIOSH, 1983), 7.

22. Kathleen Meister, *Health and Safety Aspects of Video Display Terminals: A Report by the American Council on Science and Health*, 2ed. (Summit, NJ: American Council on Science and Health, 1985), 28.

23. Gunnar Aronsson, Ing-Marie Lidström, and Rolf Meyerhoff, "Computerization and Terminal Work—Influences on Man and Working Environment," in *Proceedings of the International Scientific Conference*, 311-313.

24. Dy, *Visual Display Units*, 67-69.

25. Michael J. Smith, "Job Stress and VDUs: Is the Technology a Problem?" in *Proceedings of the International Scientific Conference*, 189-195.

26. James S. House, "Barriers to Work Stress: I. Social Support," in *Behavioral Medicine*, 157-180.

27. Robert D. Caplan et al., *Job Demands and Worker Health: Main Effects and Occupational Differences* (Ann Arbor: University of Michigan, 1980), 197.

28. Meister, *Health and Safety Aspects*, 28-29.

29. Charles J. De Wolff, "Stress Intervention at the Organizational Level," in *Behavioral Medicine*, 245.

30. U.S. Department of Health and Human Services, *Job Stress Factors*, 14-15.

31. Gould and Grischkowsky, "Doing the Same Work," 336.

32. Gunnar Aronsson, "Work Content, Stress and Health in Computer-mediated Work: A Seven Year Follow-Up Study," in *Work with Display Units 86*, 732-738. The preliminary analysis did, however, find that "the VDT-work load index is positively related to levels of adrenaline."

33. "VDT Health Debate," *Office Systems Ergonomics Report* 5(4):18-20 (1986).

34. "Follow-Ups: Heart Study," *VDT News* 4(4):4 (1987).

35. Bonnie C. Seamonds and Clinton G. Weiman, "Health Promotion in Automated Offices—High Tech Stress and the Impact of Video Display Terminals in the Workplace," in *Proceedings of the International Scientific Conference*, 394.

36. Waldemar Karwowski and Anil Mital, "Performance Standards and Job Stress in VDU Work," in *Proceedings of the International Scientific Conference*, 90; Cary L. Cooper, "The Stress of Work: An Overview," *Aviation, Space, and Environmental Medicine* 56(7):627 (1985); Smith, Carayon, and Miezo, "VDT Technology," 695-712.

37. Seamonds and Weiman, "Health Promotion in Automated Offices," 394.

38. Raija Kalimo and Anneli Leppänen, "Feedback from Video Display Terminals, Performance Control and Stress in Text Preparation in the Printing Industry," *Journal of Occupational Psychology* 58:38 (1985).

39. Michael J. Smith, "The Physical, Mental, and Emotional Stress Effects of VDT Work," *IEEE Computer Graphics and Applications* 4(4):26 (1984).

40. Arye R. Ephrath, "On Strong Backs and Weak Minds: The Leontief Dilemma," in *Proceedings of the International Scientific Conference*, 592.

41. Meister, *Health and Safety Aspects*, 28.

42. Jeanne M. Stellman et al., "Comparison of Well-Being among Non-Machine Interactive Clerical Workers and Full-Time and Part-Time VDT Users and Typists, [a,b]," in *Work with Display Units 86*, 605-614.

43. Frank Pot, Alfred Brouwers, and Pieter Padmos, "Determinants of the VDU Operator's Well-Being. 2. Workorganization," in *Proceedings of the International Scientific Conference*, 323.

44. M. Franz Schneider, "The Relationship between Ergonomics and Office Productivity," *Office Ergonomics Review* 1(1):13 (1984).

45. Dy, *Visual Display Units*, 1-14.

46. Meister, *Health and Safety Aspects*, 28-29.

47. E.N. Corlett, "Problems of Work Organization under Conditions of Technological Change," in *Changes in Working Life*, ed. K.D. Duncan, M.M. Gruneberg, and D. Wallis (New York: John Wiley, 1980), 22.

48. Brown, Dismukes, and Rinalducci, "Video Display Terminals and Vision of Workers," 133.

49. Ronald Goodrich, "Seven Office Evaluations: A Review," *Environment and Behavior* 14(3):357-358 (1982).

50. Mark A. Lieberman, Gad J. Selig, and John J. Walsh, *Office Automation: A Manager's Guide for Improved Productivity* (New York: John Wiley, 1982), 154.

51. Schneider, "The Relationship between Ergonomics and Office Productivity," 14.

52. Svend-Erik Hermansen, "Total Effects by Visual Display Unit Work," in *Proceedings of the International Scientific Conference*, 106.

53. Marilyn S. Joyce, "Ergonomic Skills Training: A Practical Approach to Improving Productivity and Worker Health," in *Proceedings of the International Scientific Conference*, 761.

54. Robin Baker and Laura Stock, "Worker Education and User Participation in the Development of Protective Policies for VDT Operators," in *Work with Display Units 86*, 665-669.

55. Niels Diffrient, "Interview: Designer Niels Diffrient Speaks Out on Ergonomics, Office Productivity and Technology," *Office Ergonomics Review* 1(1):11 (1984).

56. Dainoff and Mark, "Task and the Adjustment of Ergonomic Chairs," 294-302.

57. "Current Trends," *Morbidity & Mortality Weekly*, 619.

58. Meister, *Health and Safety Aspects*, 28.

59. J.J. Allegro, "An Action Research Project on Office Automation," in *Proceedings of the International Scientific Conference*, 755-758.

60. T.F.M. Stewart, "Software Ergonomics," in *Ergonomics and Health in Modern Offices*, ed. Etienne Grandjean (London: Taylor & Francis, 1984), 153-159.

61. Pot, Brouwers, and Padmos, "Determinants of the VDU Operator's Well-Being," 323.

62. Ibid.

63. American Medical Association Council on Scientific Affairs, "Health Effects of Video Display Terminals: Council Report," *Journal of the American Medical Association* 257(11):1510 (1987); U.S. Department of Health and Human Services, *Job Stress Factors*, 12-15; Meister, *Health and Safety Aspects*, 28-29.

64. R.H. Irving, C.A. Higgins, and F.R. Safayeni, "Computerized Performance Monitoring Systems: Use and Abuse," *Communications of the ACM* 29(8):800 (1986).

65. U.S. Department of Health and Human Services, National Institute for Occupational Safety and Health, *Potential Health Hazards of Video Display Terminals* (Cincinnati, OH: NIOSH, 1981), 69-70.

66. Dy, *Visual Display Units*, 113-125.

67. Hewlett-Packard, *Productivity through Ergonomically Designed Workstations: A Quality Team Study of the Effects of Eye and Body Strain on Video Display Terminal Users* (Internal Report) (n.p.: Hewlett-Packard, 1984), 14.

68. R. Mallette, *A Supervisor's Guide to Video Display Work*, 21-22.

69. Peter Unterweger, "Appropriate Automation: Thoughts on Swedish Examples of Sociotechnical Innovation," *Labor Law Journal* 36(8):570 (1985).

Notes to Chapter 6: *Pregnancy and Reproduction*

1. Arthur D. Bloom, ed., *Guidelines for Studies of Human Populations Exposed to Mutagenic and Reproductive Hazards* (White Plains, NY: March of Dimes Birth Defects Foundation, 1981), 77, 47.

2. "Spontaneous Abortions and Work," *WOHRC News* (Women's Occupational Health Resource Center) 8(2):3 (1987).

3. Bloom, *Guidelines for Studies of Human Populations*, 47, 71-74, 79.

4. The interested reader may learn more about the stages of embryo development, the types of abnormality associated with abnormal cell division at various stages, and agents known to cause such abnormal development in *The Silent Intruder: Surviving the Radiation Age*, by Charles Panati and Michael Hudson (Houghton-Mifflin, 1981). Despite its sensational title, this book provides a fairly well-reasoned discussion of radiation, recognizing not only its risks but also its benefits.

5. Much of this chronology was compiled from reports in *VDT News* and *Office Systems Ergonomics Report.*

6. U.S. Department of Health and Human Services, "Health Hazard Evaluation Report, Southern Bell, Atlanta, Georgia," HETA 83-329-1498 (Cincinnati: National Institute for Occupational Safety and Health, 1984).

7. U.S. Department of Health and Human Services, National Centers for Disease Control, *Cluster of Spontaneous Abortions*, EPI-80-113-2 (Atlanta: National Centers for Disease Control, 1981).

8. *The Office Systems Ergonomics Report* 2(6):2-3 (1983).

9. Angell, *Health and Safety Effects*, 34-35. According to Angell, "The whole study was so experimentally unsound that its results are rendered virtually meaningless."

10. U.S. Department of Health and Human Services, "Health Hazard Evaluation, United Airlines, San Francisco, California," HETA 84-191 (Cincinnati: National Institute for Occupational Safety and Health, 1985).

11. DeMatteo, *Terminal Shock*, 27.

12. William Halperin, "NIOSH Retrospective Study of the Impact of VDT Use on Pregnancies," (speech delivered at the Second International Information Industry Conference, May 1986, Dallas, TX), 3.

13. Halperin, "NIOSH Retrospective Study," 2.

14. "If the expected incidence of an outcome in an unexposed population is 0.001, and the risk of the outcome is doubled in exposed individuals, then a population of 10,100 exposed individuals must be studied in order to have a 75 percent probability of demonstrating a difference significant at the 5 percent level (using a two-tailed test). In this calculation we have assumed that the expected incidence is constant and that data on an unexposed sample will not be needed. If this assumption cannot be made and two samples (exposed and unexposed) are required, then one would need 20,195 individuals in each sample to detect a relative risk of 2.0." (Bloom, *Guidelines for Studies of Human Populations*, 89.)

15. Colin Mackay, "The Alleged Reproductive Hazards of VDUs," in *Work & Stress* 1 (1):50 (1987).

16. Becker and Selden, *The Body Electric*, 276-288.

Notes to Chapter 7: *Policy and Regulatory Issues*

1. Kenneth R. Foster, "The VDT Debate," *American Scientist* 74(Mar/Apr):163-168 (1986).

2. "Special Issue on Visual Display Units," *Conditions of Work Digest* 5(1):1986.

3. David LeGrande and David J. Eisen, "Developing a Union VDU Ergonomic Strategy," in *Proceedings of the International Scientific Conference*, 130-133.

4. New Mexico, Office of the Governor, "State of New Mexico Executive Order Covering Use of Video Display Terminals by State Employees" (Santa Fe: March 1985).

5. Rosemary Guiley, "Business, Labor Clash over Regulating Computer Terminals in the Workplace," *Management Review*, December 1984, 38.

6. Arndt and Chapman, *Potential Office Hazards and Controls*, 114.

7. Testimony before the House of Representatives Subcommittee on Health and Safety, OSHA Oversight Committee on Video Displays in the Workplace, June 12, 1984.

8. "DPMA Opposes VDT Bills," *Displays* 7(2):95 (1986).

9. T.J. Springer, "The Statutes and Standards Movement: Smoke before Fire?" *Office Ergonomics Review* 2(2):14-15 (1985).

10. T.F.M. Stewart, "Interview: From the Beginning to the End, Performance Drives Solutions," *The Report Store Report & Catalogue,* Winter 1987, 6-7.

11. Paula S. Bellis, "Friend or Foe? The VDT Controversy Continues," *The Office* 103(2):41-42 (February 1986).

12. Alan F. Westin et al., *The Changing Workplace: A Guide to Managing the People, Organizational and Regulatory Aspects of Office Technology* (White Plains, NY: Knowledge Industry Publications, 1985), Section 6-2.

13. Mallette, *A Supervisor's Guide to Video Display Work*, 18.

14. Guiley, "Business, Labor Clash," 38.

15. Westin et al., *The Changing Workplace*, Section 6-2.

16. Ibid., Section 9-3.

17. Ibid., Section 9-14.

18. Bellis, "Friend or Foe?" 41.

19. Ibid.

20. Westin et al., *The Changing Workplace*, Section 8-7.

21. Gary Dessler, "Unions Win Benefits for Users of VDTs," *Kansas City Star*, 7 September 1986, sec. E, p.6.

22. National Council on Compensation Insurance, *VDTs and Office Safety: An Overview* (New York: NCCI Information Products, 1986), 23-25.

23. U.S. Congress, Office of Technology Assessment, *Automation of America's Offices, 1985-2000* (Washington: U.S. Government Printing Office, 1985), n.139.

24. "Current Trends," *Morbidity & Mortality Weekly Report*, 613-614.

25. Leonard A. Sagan, "Problems in Health Measurement for the Risk Assessor," in *Technological Risk Assessment*, ed. Paolo F. Ricci, Leonard A. Sagan, and Chris G. Whipple (Norwell, MA: Martinus Nijhoff, 1984), 1-29.

26. Ibid.

27. M. Franz Schneider, "Why Ergonomics Can No Longer Be Ignored," *Office Administration and Automation*, July 1985, 29.

28. Noe Palacios, "Planning the Office of the Future—Today," in *Proceedings of the International Scientific Conference*, 480.

29. Westin et al., *The Changing Workplace*, Section 7-8.

30. Robert M. Kaplan, Michael H. Criqui, eds., *Behavioral Epidemiology and Disease Prevention* (New York: Plenum, 1985), vii.

List of Works Cited

Adey, W. Ross. "The Energy Around Us." *The Sciences* 26(1):53-58 (1986).

Allegro, J.J. "An Action Research Project on Office Automation." In *Proceedings of the International Scientific Conference*, 755-758. *See* Swedish National Board 1986.

American Industrial Hygiene Association. *Health and Ergonomic Considerations of Visual Display Units—Symposium Proceedings*. Akron, OH: American Industrial Hygiene Association, 1983.

American Medical Association. Council on Scientific Affairs. "Health Effects of Video Display Terminals: Council Report." *Journal of the American Medical Association* 257(11):1508-1512 (1987).

Angell, Carolyn. *The Health and Safety Effects of Visual Display Units—A Review*. London: The Technical Change Centre, 1986.

"Arbitrator Finds No Health Hazard in Use of Video Display Terminals." *American Newspaper Publishers Association, Labor & Personnel Relations Bulletin* no. 6496:77-83 (3 March 1978).

Armstrong, Thomas J., Robert G. Radwin, Doan J. Hansen, and Kenneth W. Kennedy. "Repetitive Trauma Disorders: Job Evaluation and Design." *Human Factors* 28(3):325-336 (1986).

Arndt, Robert. "Body Posture." In *Health and Ergonomic Considerations of Visual Display Units*, 29-44. *See* American Industrial Hygiene Association 1983.

Arndt, Robert. "Working Posture and Musculoskeletal Problems of Video Display Terminal Operators—Review and Reappraisal." *American Industrial Hygiene Association Journal* 44(6):437-446 (1983).

Arndt, Robert, and Larry Chapman. *Potential Office Hazards and Controls. Contractor's Report Prepared for the Office of Technology Assessment*. Madison: University of Wisconsin, 1984.

Aronsson, Gunnar. "Work Content, Stress and Health in Computer-Mediated Work: A Seven Year Follow-Up Study." In *Work with Display Units 86*, 732-738. *See* Knave 1987.

Aronsson, Gunnar, Ing-Marie Lidström, and Rolf Meyerhoff. "Computerization and Terminal Work—Influences on Man and Working Environment." In *Proceedings of the International Scientific Conference*, 311-313. *See* Swedish National Board 1986.

Australian National University. *ANU Research on RSI: Proceedings*. Canberra: Australian National University, 1986.

Bachman, W., J. Kupfer, B. Hinz, and D. Methling. "Occupational Health Aspects of Work with Display Units." In *Proceedings of the International Scientific Conference*, 983-986. *See* Swedish National Board 1986.

Baker, Robin, and Laura Stock. "Worker Education and User Participation in the Development of Protective Policies for VDT Operators." In *Work with Display Units 86*, 665-669. *See* Knave 1987.

Bammer, Gabriele. "Musculo-Skeletal Problems Associated with VDU Use at the Australian National University: A Case Study of Changes in Work Practices." In *Trends in Ergonomics/Human Factors III: Proceedings of the Annual International Industrial Ergonomics and Safety Conference*, 285-293. *See* Karwowski 1986.

Bammer, Gabriele, and Ilse Blignault. "A Review of Research on RSI." *ANU Research on RSI: Proceedings*, 5-9. *See* Australian National University 1986.

Barresi, Barry G., and Jesse Rosenthal. *New York State Occupational Vision Benefit Plan Study: An Evaluation of a Vision Plan for VDT Users and Office Workers*. New York: State University of New York, Center for Vision Care Policy, 1986.

Becker, Robert O., M.D., and Andrew A. Marino. *Electromagnetism and Life*. Albany: State University of New York Press, 1982.

Becker, Robert O., M.D., and Gary Selden. *The Body Electric: Electromagnetism and the Foundation of Life*. New York: William Morrow, 1985.

Bellis, Paula S. "Friend or Foe? The VDT Controversy Continues." *The Office* 103(2):41-42 (1986).

Bennett, John, Donald Case, Jon Sandelin, and Michael J. Smith. *Visual Display Terminals: Usability Issues and Health Concerns*. Englewood Cliffs, NJ: Prentice-Hall, 1984.

Bergqvist, Ulf OV. "Video Display Terminals and Health: A Technical and Medical Appraisal of the State of the Art." *Scandinavian Journal of Work, Environment & Health* 10(Supplement 2):(1984).

Bjørset, Hans-Henrick. "Lighting for Visual Display Unit Workplaces." In *Work with Display Units 86*, 331-339. *See* Knave 1987.

Blau, Rachel. "Repetitive Strain Injury: A New High Tech Epidemic." *Video Views* 3(2) (1986).

Bloom, Arthur D., ed. *Guidelines for Studies of Human Populations Exposed to Mutagenic and Reproductive Hazards*. White Plains, NY: March of Dimes Birth Defects Foundation, 1981.

Bonnell, J.A. "VDTs and Health: Fact or Fancy?" In *Work with Display Units 86*, 3-5. *See* Knave 1987.

Boyce, P.R. *Human Factors in Lighting*. New York: MacMillan, 1981.

Brown, B.S., K. Dismukes, and E.J. Rinalducci. "Video Display Terminals and Vision of Workers: Summary and Overview of a Symposium." *Behaviour and Information Technology* 1(2):121-140 (1982).

Çakir, A. "Are VDTs a Health Hazard?" Paper presented at the World Conference on Ergonomics in Computer Systems, Los Angeles, September 1986.

Çakir, A., D.J. Hart, and T.F.M. Stewart. *Visual Display Terminals: A Manual Covering Ergonomics, Workplace Design, Health and Safety, Task Organization*. New York: John Wiley and Sons, 1980.

Campbell, F.W., and K. Durden. "The Visual Display Terminal Issue: A Consideration of Its Physiological, Psychological and Clinical Background." *Ophthalmic and Physiological Optics* 3(2):175-192 (1983).

Caplan, Robert D., Sidney Cobb, John R.P. French, Jr., R. Van Harrison, and S.R. Pinneau, Jr. *Job Demands and Worker Health*. Ann Arbor: University of Michigan, 1980.

Cohen, Sheldon, Gary W. Evans, Daniel Stokols, and David S. Krantz. *Behavior, Health, and Environmental Stress*. New York: Plenum, 1986.

"Concern Grows over RSIs." *VDT News* 4(1):2-3 (1987).

Connecticut. General Assembly. General Law Committee. "VDT Radiation," a special report by the Connecticut Academy of Science and Engineering. Hartford: State of Connecticut, 1984 (SA83-54).

Cooper, Cary L. "The Stress of Work: An Overview." *Aviation, Space, and Environmental Medicine* 56(7):627-632 (1985).

Corlett, E.N. "Problems of Work Organization under Conditions of Technological Change." In *Changes in Working Life*, 17-30. *See* Duncan 1980.

Corlett, E.N. and J. Richardson, eds. *Stress, Work Design, and Productivity*. New York: John Wiley and Sons, 1981.

Cox, Tom. "The Nature and Measurement of Stress." *Ergonomics* 28(8):1155-1163 (1985).

Cox, Tom and Colin Mackay. "A Transactional Approach to Occupational Stress." In *Stress, Work Design, and Productivity*, 91-113. *See* Corlett 1981.

"Current Trends: Leading Work-Related Disease and Injuries." *Morbidity & Mortality Weekly Report* 35(39):613-621 (1986).

Dainoff, Marvin J., and Marilyn Hecht Dainoff. *People & Productivity: A Manager's Guide to Ergonomics in the Electronic Office.* Toronto: Holt, Rinehart & Winston of Canada, 1986.

Dainoff, Marvin J., and Leonard S. Mark. "Task and the Adjustment of Ergonomic Chairs." In *Work with Display Units 86,* 294-302. *See* Knave 1987.

Damodaran, Leela. "Introduction." In *Health Hazards of VDTs?* 3-4. *See* Pearce 1984.

Davis, Bill. "Carpal Tunnel Syndrome Strikes VDT Operators." *Video Views* 2(3):3 (1985).

de Groot, Jan, and Andre Kamphuis. "Eyestrain in VDU Users: Physical Correlates and Long-Term Effects." *Human Factors* 25(4):409-415 (1983).

"DEMA Survey Cites Problems among Data Entry Operators." *The Office Health & Safety Monitor* 1(1):3 (1986).

DeMatteo, Bob. *Terminal Shock.* Toronto: NC Press Limited, 1985.

Dessler, Gary. "Unions Win Benefits for Users of VDTs." *Kansas City Star,* 7 September 1986, Section E.

de Wolff, Charles J. "Stress and Strain in the Work Environment: Does It Lead to Illness?" In *Behavioral Medicine,* 33-43. *See* Gentry 1985.

de Wolff, Charles J. "Stress Intervention at the Organizational Level." In *Behavioral Medicine,* 241-252. *See* Gentry 1985.

Dickerson, O. Bruce, and Walter E. Baker. "Health Considerations at the Information Workplace." In *Visual Display Terminals,* 271-286. *See* Bennett 1984.

Diffrient, Niels. "Interview: Designer Niels Diffrient Speaks Out on Ergonomics, Office Productivity and Technology." *Office Ergonomics Review* 1(1):10-11 (1984).

"DPMA Opposes VDT Bills." *Displays* 7(2):95 (1986).

Duncan, K.D., M.M. Gruneberg, and D. Wallis, eds. *Changes in Working Life: Proceedings of an International Conference on Changes in the Nature and Quality of Working Life.* New York: John Wiley and Sons, 1980.

Dy, Fe Josefina F. *Visual Display Units: Job Content and Stress in Office Work.* Geneva: International Labour Office, 1985.

Edström, Ricardo. "Regulating VDU Work Places." In *Proceedings of the International Scientific Conference*, 987-988. *See* Swedish National Board 1986.

"EPA Ready to Ship Equipment to Labs for Delgado-Like Tests." *COT Center Report* 2(16):2 (1986).

Ephrath, Arye P. "On Strong Backs and Weak Minds: The Leontief Dilemma." In *Proceedings of the International Scientific Conference*, 590-592. *See* Swedish National Board 1986.

"Ergonomics of Office Keyboards." *Office Systems Ergonomics Report* 5(2): (1986).

Ericson, Anders, and Bengt Källen. "An Epidemiological Study of Work with Video Screens and Pregnancy Outcome: II. A Case-control Study." *American Journal of Industrial Medicine* 9(5):459-475 (1986).

"Et Cetera." *Office Systems Ergonomics Report* 5(5):16 (1986).

European Computer Manufacturers Association. *Visual Displays: Health Aspects*. Geneva: ECMA, 1985.

Evans, Jennifer. "Questionnaire Survey of British VDU Operators." In *Proceedings of the International Scientific Conference*, 565-569. *See* Swedish National Board 1986.

Feeley, Helen. *The VDT Operator's Problem Solver*. Ottawa, Ontario: Planetary Association for Clean Energy, 1984.

Fertig, Robert T. "VDT/Miscarriage Link Seen in Quebec Survey." *Management Information Systems Week* 6(23) (1985).

"Follow-Ups: Heart Study." *VDT News* 4(4):4 (1987).

Foster, Kenneth R. "The VDT Debate." *American Scientist* 74:163-168 (March/ April 1986).

Galitz, Wilbert O. "Proper Planning Stems Automation Atrophy." *National Safety and Health News*, October 1985, 96-99.

Gentry, W. Doyle, Herbert Benson, and Charles J. De Wolff, eds. *Behavioral Medicine: Work, Stress and Health*. Norwell, MA: Martinus Nijhoff, 1985.

"Glare." *Office Systems Ergonomics Report* 4(1): (1985).

Goodrich, Ronald. "Seven Office Evaluations: A Review." *Environment and Behavior* 14(3):353-378 (1982).

Gould, John D., and Nancy Grischkowsky. "Doing the Same Work with Hard Copy and with Cathode-Ray Tube (CRT) Computer Terminals." *Human Factors* 26(3):323-357 (1984).

Grandjean, Etienne. *Ergonomics in Computerized Offices.* London: Taylor & Francis, 1987.

Grandjean, Etienne, ed. *Ergonomics and Health in Modern Offices.* London: Taylor & Francis, 1984.

Grandjean, Etienne, W. Hünting, and M. Pidermann. "VDT Workstation Design: Preferred Settings and Their Effects." *Human Factors* 25(2):161-175 (1983).

Grandjean, Etienne, and E. Vigliani, eds. *Ergonomic Aspects of Visual Display Terminals.* London: Taylor & Francis, 1980.

Guiley, Rosemary. "Business, Labor Clash over Regulating Computer Terminals in the Workplace." *Management Review,* December 1984, 37-39.

Halperin, William. "NIOSH Retrospective Study of the Impact of VDT Use on Pregnancies." Speech delivered at the Second International Information Industry Conference, May 1986, Dallas, TX.

Hansson, Jan-Erik, and Monika A. Attebrant. "The Effect of Table Height and Table Top Angle on Head Position and Reading Distance." In *Proceedings of the International Scientific Conference*, 419-422. *See* Swedish National Board 1986.

Hedman, Leif R., and Valdimar Bhiem. "Short-Term Changes in Eyestrain of VDU Users as a Function of Age." *Human Factors* 26(3):357-370 (1984).

Helander, Martin G., Patricia A. Billingsley, and Jayne M. Schurick. "An Evaluation of Human Factors Research on Visual Display Terminals in the Workplace." In *Human Factors Review 1984*, 55-129. *See* Muckler 1984.

Hembree, Diana, and Sarah Henry. "A Newsroom Hazard Called RSI." *Columbia Journalism Review* 25:19-24 (Jan/Feb 1987).

Henriques, Vico. "Prepared statement of Vico Henriques . . . June 12, 1984," 297-358. *See* U.S. Congress 1984.

Hermansen, Svend-Erik. "Total Effects by Visual Display Unit Work." In *Proceedings of the International Scientific Conference*, 105-107. *See* Swedish National Board 1986.

Hewlett-Packard. *Productivity through Ergonomically Designed Workstations: A Quality Team Study of the Effects of Eye and Body Strain on Video Display Terminal Users (Internal Report).* Hewlett-Packard, 1984.

Hopper, Grace. Interview by Morley Safer. "60 Minutes" Transcript 28(50):6-11 (24 August 1986).

Horwitz, Tony, and Geraldine Brooks. "Chronicle: 'Buggered for Life'—by VDTs." *Columbia Journalism Review* 24(2):10-12 (1985).

House, James S. "Barriers to Work Stress: I. Social Support." In *Behavioral Medicine*, 157-180. *See* Gentry 1985.

Howarth, Peter Alan, and Howell Owen Istance. "The Association Between Visual Discomfort and the Use of Visual Display Units." *Behaviour and Information Technology* 4(2):131-149 (1985).

Howarth, Peter Alan, and Howell Owen Istance. "The Validity of Subjective Reports of Visual Discomfort." *Human Factors* 28(3):347-351 (1986).

Human Factors Society. *American National Standard for Human Factors Engineering of Visual Display Terminal Workstations: Revised Review Draft.* Santa Monica, CA: Human Factors Society, 1986.

Institute de recherche en santé et en sécurité du travail du Québec. *Report of the Task Force on Video Display Terminals and Workers' Health.* Montreal: IRSST, 1985.

International Commission on Illumination. *Vision and the Visual Display Unit Workstation.* Paris: CIE, 1984.

Irving, R.H., C.A. Higgins, and F.R. Safayeni. "Computerized Performance Monitoring Systems: Use and Abuse." *Communications of the ACM* 29(8):794-801 (1986).

Iwasaki, Tsuneto, and Shinji Kurimoto. "Objective Evaluation of Eye Strain using Measurements of Accommodative Oscillation." *Ergonomics* 30(3):581-587 (1987).

Joyce, Marilyn S. "Ergonomic Skills Training: A Practical Approach to Improving Productivity and Worker Health." In *Proceedings of the International Scientific Conference*, 759-762. *See* Swedish National Board 1986.

Kalimo, Raija, and Anneli Leppänen. "Feedback from Video Display Terminals, Performance Control and Stress in Text Preparation in the Printing Industry." *Journal of Occupational Psychology* 58:27-38 (1985).

Kaplan, Robert M., and Michael H. Criqui, eds. *Behavioral Epidemiology and Disease Prevention.* New York: Plenum, 1985.

Kariya, Scott. "Seeking a Perfect Chair." *PC Magazine*, 2 October 1984, 140-148.

Karwowski, Waldemar, ed. *Trends in Ergonomics/Human Factors III: Proceedings of the Annual International Industrial Ergonomics and Safety Conference.* Amsterdam: North-Holland, 1986.

Karwowski, Waldemar, and Anil Mital. "Performance Standards and Job Stress in VDU Work." In *Proceedings of the International Scientific Conference,* 90-93. *See* Swedish National Board 1986.

Knave, Bengt, and Per-Gunnar Widebäck, eds. *Work with Display Units 86.* Amsterdam: North-Holland, 1987.

Landsberger, H.A. *Hawthorne Revisited: Management and the Worker, Its Critics and Developments in Human Relations in Industry.* Ithaca, NY: Cornell University Press, 1958.

Läubli, Thomas. "Review on Working Conditions and Postural Discomfort in VDT Work." In *Proceedings of the International Scientific Conference,* 3-6. *See* Swedish National Board 1986.

"Lead Aprons for Operators: Help or Hindrance?" *VDT News* 1(4):16 (1984).

LeGrande, David, and David J. Eisen. "Developing a Union VDU Ergonomic Strategy." In *Proceedings of the International Scientific Conference,* 131-133. *See* Swedish National Board 1986.

Levy, Finn, and Ingrid Greger Ramberg. "Eye Fatigue among VDU Users and Non-VDU Users." In *Work with Display Units 86,* 42-52. *See* Knave 1987.

Lidén, Carola, and Jan E. Wahlberg. "Does Visual Display Terminal Work Provoke Rosacea?" *Contact Dermatitis* 13:235-241 (1985).

Lieberman, Mark A., Gad J. Selig, and John J. Walsh. *Office Automation: A Manager's Guide for Improved Productivity.* New York: John Wiley and Sons, 1982.

Lindén, Victor, and Sturla Rolfsen. "Video Computer Terminals and Occupational Dermatitis," (letter to the editor). *Scandinavian Journal of Work, Environment & Health* 7:62-64 (1981).

Lindorff, Dave. "Frontlines: Epidemic of Stiff Little Fingers." *Mother Jones,* May 1985, 11.

Mackay, Colin. "The Alleged Reproductive Hazards of VDUs." *Work & Stress* 1(1):49-57 (1987).

"Major Unions Mapping Strategy on Occupational Stress Issue." *The Office Health & Safety Monitor* 1(1):1-2 (1986).

Mallette, R. *A Supervisor's Guide to Video Display Work.* Toronto: Ontario Hydro, 1984.

Mandal, A.C. *The Seated Man: Homo Sedens.* Denmark: Dafnia, 1985.

Meister, Kathleen A. *Health and Safety Aspects of Video Display Terminals: A Report by the American Council on Science and Health.* 2ed. Summit, NJ: American Council on Science and Health, 1985.

Mikøajczyk, H., J. Indulski, T. Kameduøa, and M. Pawlaczyk. "Effects of TV Sets Electromagnetic Fields on Rats." In *Work with Display Units 86,* 122-128. *See* Knave 1987.

Mikøajczyk, H., J. Indulski, T. Kameduøa, and M. Pawlaczyk. "Task-Load and Endocrinological Risk for Pregnancy in Women VDU Operators." In *Work with Display Units 86,* 115-121. *See* Knave 1987.

Miller, Stephen C. *The Eye Care Book for Computer Users.* St. Louis: Eye Care Concepts, 1986.

Monat, Alan, and Richard S. Lazarus. *Stress and Coping: An Anthology.* New York: Columbia University Press, 1985.

Muckler, Frederick A., ed. *Human Factors Review 1984.* Santa Monica, CA: Human Factors Society, 1984.

Murray, William E. "Is There a Radiation Risk?" In *Health and Ergonomic Considerations of Visual Display Units,* 87-101. *See* American Industrial Hygiene Association 1983.

National Council on Compensation Insurance. *VDTs and Office Safety: An Overview.* New York: NCCI Information Products, 1986.

National Research Council. Committee on Vision. *Video Displays, Work, and Vision.* Washington: National Academy Press, 1983.

New Mexico. Office of the Governor. "State of New Mexico Executive Order covering Use of Video Display Terminals by State Employees." Santa Fe: Office of the Governor, March 1985.

"The 9-to-5 Report on VDT Health Hazards," *Office Systems Ergonomics Report* 2(6):1-5 (1983).

"Non-Ionizing Radiation Biohazards and VDTs." *Office Systems Ergonomics Report* 3(5):8-11 (1984).

"Numbers Worth Knowing." *VDT News* 4(3):2 (1987).

"NY Workers' Compensation Board Rules Cataracts Caused by VDTs." *VDT News* 2(1):7-8 (1985).

Nyman, Karla Gösta. "Refraction in VDU Operators - A Comparison with Other Professions." In *Work with Display Units 86,* 556-561. *See* Knave 1987.

Olsen, Walter Cato. "Facial Particle Exposure in the VDU Environment: The Role of Static Electricity." In *Proceedings of the International Scientific Conference*, 797-800. *See* Swedish National Board 1986.

Ong, Choon-Nam. "VDT Work Place Design and Physical Fatigue: A Case Study in Singapore." In *Ergonomics and Health in Modern Offices*, 484-494. *See* Grandjean 1984.

Ong, Choon-Nam, and Wai-on Phoon. "Influence of Age on Performance and Health of VDT Workers." In *Work with Display Units 86*, 211-215. *See* Knave 1987.

"ONR PEMF Project Delayed." *Microwave News: A Report on Non-Ionizing Radiation* 6(3):11 (1986).

Palacios, Noe. "Planning the Office of the Future—Today." In *Proceedings of the International Scientific Conference*, 477-480. *See* Swedish National Board 1986.

Panati, Charles, and Michael Hudson. *The Silent Intruder: Surviving the Radiation Age*. Boston: Houghton-Mifflin, 1981.

Pearce, Brian, ed. *Health Hazards of VDTs?* New York: John Wiley and Sons, 1984.

"PEMFs and Chick Eggs: Results Mixed, Mechanism Elusive." *Microwave News: A Report on Non-Ionizing Radiation* 6(3) (1986).

Pollack, Herbert, M.D. "Epidemiologic Data on American Personnel in the Moscow Embassy." *Bulletin of the New York Academy of Medicine* 55(11):1182-1186 (1979).

Pot, Frank, Alfred Brouwers, and Pieter Padmos. "Determinants of the VDT Operator's Well-Being. 2. Workorganization." In *Proceedings of the International Scientific Conference*, 322-324. *See* Swedish National Board 1986.

"Radiofrequency Radiation and Video Display Terminals." *Office Systems Ergonomics Report* 3(5) (1984).

"Ramazzini's Corner: VDTs and Epilepsy." *At the Centre* 9(6):17 (1987).

Rand, George. "Caution: The Office Environment May Be Hazardous to Your Health." *AIA Journal* 68(12) (1979).

Ricci, Paolo F., Leonard A. Sagan, and Chris G. Whipple, eds. *Technological Risk Assessment*. Norwell, MA: Martinus Nijhoff, 1984.

Rose, Lawrence, M.D. "Workplace Video Display Terminals and Visual Fatigue." *Journal of Occupational Medicine* 29(4):321-324 (1987).

Rowe, Susan, Maurice Oxenburgh, and David Douglas. "Repetition Strain Injury in Australian VDU Users." In *Work with Display Units 86*, 38-41. *See* Knave 1987.

Rupp, Bruce, George Mine, M.L. Wolbarsht, and Ian Andrew. *Human Factors of Workstations with Visual Displays*. 3ed. San Jose, CA: International Business Machines, 1984.

Sagan, Leonard A. "Problems in Health Measurement for the Risk Assessor." In *Technological Risk Assessment*, 1-29. *See* Ricci 1984.

Sauter, Steven, L., L. John Chapman, and Sheri J. Knutson. *Improving VDT Work: Causes and Control of Health Concerns in VDT Use*. Lawrence, KS: The Report Store, 1986.

Sauter, Steven, L., L. John Chapman, Sheri J. Knutson, and Henry A. Anderson. "Wrist Trauma in VDT Keyboard Use: Evidence, Mechanisms and Implications for Keyboard and Wristrest Design." In *Proceedings of the International Scientific Conference*, 230-233. *See* Swedish National Board 1986.

Sauter, Steven L., Peter L. Eichman, L. John Chapman, Sheri J. Knutson, Henry A. Anderson, and Robert H. Arndt. "Chronic Neck-Shoulder Discomfort in VDT Use: Prevalence and Medical Observations." In *Proceedings of the International Scientific Conference*, 154-157. *See* Swedish National Board 1986.

Schleifer, Lawrence M., and Steven L. Sauter. "Controlling Glare Problems in the VDT Work Environment." *Library Hi Tech* 3(4):21-25 (1985).

Schneider, M. Franz. "The Relationship between Ergonomics and Office Productivity." *Office Ergonomics Review* 1(1):12-15 (1984).

Schneider, M. Franz. "Why Ergonomics Can No Longer Be Ignored." *Office Administration and Automation* 46(7):26-29 (1985).

Seamonds, Bonnie C., and Clinton G. Weiman. "Health Promotion in Automated Offices—High Tech Stress and the Impact of Video Display Terminals in the Workplace." In *Proceedings of the International Scientific Conference*, 391-394. *See* Swedish National Board 1986.

Selye, Hans. "History and Present Status of the Stress Concept." In *Stress and Coping: An Anthology*, 17-29. *See* Monat 1985.

Shroder-Frechette, K.S. *Risk Analysis and Scientific Method*. Boston: D. Reidel, 1985.

Slesin, Louis. "People Are Antennas, Too: The Biology of the Electromagnetic Spectrum." *Whole Earth Review*, Spring 1986, 50-55.

Sliney, D.H., M.L. Wolbarsht, and A.M. Muc. "Editorial: Differing Radiofrequency Standards in the Microwave Region— Implications for Future Research." *Health Physics* 49(5):677-683 (1985).

Smith, Michael J. "Job Stress and VDUs: Is the Technology a Problem?" In *Proceedings of the International Scientific Conference*, 189-195. *See* Swedish National Board 1986.

Smith, Michael J. "The Physical, Mental, and Emotional Stress Effects of VDT Work." *IEEE Computer Graphics and Applications* (4)4:23-27 (1984).

Smith, Michael J., Pascale Carayon, and Kathleen Miezio. "VDT Technology: Psychosocial and Stress Concerns." In *Work with Display Units 86*, 695-712. *See* Knave 1987.

"Special Issue on Visual Display Units." *Conditions of Work Digest* 5(1): 1986.

Spengler, John D., and Ken Sexton. "Indoor Air Pollution: A Public Health Perspective." *Science* 221(4605):9-17 (1983).

"Spontaneous Abortions and Work," *WOHRC News* 8(2):3 (1987).

Springer, T.J. "The Statutes and Standards Movement: Smoke before Fire?" *Office Ergonomics Review* 2(2):14-15 (1985).

Starr, Steven J. "Effects of Video Display Terminals in a Business Office." *Human Factors* 26(3):347-356 (1984).

Stellman, Jeanne M., Susan Klitzman, Gloria R. Gordon, and Barry Snow. "Comparison of Well-Being among Non-Machine Interactive Clerical Workers and Full-Time and Part-Time VDT Users and Typists, [a,b]." In *Work with Display Units 86*, 605-614. *See* Knave 1987.

Steneck, Nicholas. *The Microwave Debate*. Cambridge: MIT Press, 1984.

Stewart, T.F.M. "Interview: From the Beginning to the End, Performance Drives Solutions," *The Report Store Report & Catalogue*, Winter 1987, 6-7.

Stewart, T.F.M. "Software Ergonomics." In *Ergonomics and Health in Modern Offices*, 153-159. *See* Grandjean 1984.

Stewart, T.F.M. "VDU Ergonomics—A Review of the Last Two Years." In *Health Hazards of VDTs?* 39-45. *See* Pearce 1984

Suess, Michael J. "Health Impact of Work with Visual Display Terminals." In *Work with Display Units 86*, 6-15. *See* Knave 1987.

"Swedish Mice Study: Effects Still Significant." *Microwave News: A Report on Non-Ionizing Radiation* 6(3):10 (1986).

Swedish National Board of Occupational Safety and Health Research
Department. *Proceedings of the International Scientific Conference: Work
with Display Units*. Stockholm: NBOSH, 1986.

Tanzer, Radford C. "The Carpal Tunnel Syndrome: A Clinical and Anatomical
Study." *The Journal of Bone and Joint Surgery* 41A:626-634 (1959).

Tarlau, Eileen Senn. "Measurement and Shielding of VLF Radiation from Video
Display Terminals." *Journal of the American Industrial Hygiene
Association* 47(1):A8-A10 (1986).

Terrana, T., F. Merluzzi, and E. Guidici. "Electromagnetic Radiations Emitted
by Visual Display Units." In *Ergonomic Aspects of Visual Display
Terminals*, 13-20. *See* Grandjean 1980.

Tibbs, Hardwin. *The Future of Light*. London: Watkins, 1981.

Tinning, R.J. and W.B. Spry. "The Extent and Significance of Stress Symptoms
in Industry—with Examples from the Steel Industry." In *Stress, Work
Design, and Productivity*, 129-148. *See* Corlett 1981.

Tjønn, Hans H. "Report of Facial Rashes among VDU Operators in Norway." In
Health Hazards of VDTs? 17-23. *See* Pearce 1984.

Trades Union Congress. *TUC Guidelines on VDUs*. London: Trades Union
Congress, 1985.

Tribukait, Bernhard, Eva Cekan, and Lars-Erik Paulsson. "Effects of Pulsed
Magnetic Fields on Embryonic Development in Mice." In *Work with Display
Units 86*, 129-131. *See* Knave 1987.

Turiel, Isaac. *Indoor Air Quality and Human Health*. Stanford, CA: Stanford
University Press, 1985.

Unterweger, Peter. "Appropriate Automation: Thoughts on Swedish Examples of
Sociotechnical Innovation." *Labor Law Journal* 36(8):570 (1985).

U.S. Congress. Office of Technology Assessment. *Automation of America's
Offices, 1985-2000*. Washington: Government Printing Office, 1985.

U.S. Congress. House. Committee on Education and Labor. Subcommittee on
Health and Safety. *OSHA Oversight—Video Display Terminals in the
Workplace* Hearings, February-June 1984. 98th Cong., 2d Sess., 1984.

U.S. Congress. House. Committee on Education and Labor. Subcommittee on
Health and Safety. *A Staff Report on the Oversight of OSHA with Respect
to Video Display Terminals in the Workplace*. 99th Cong., 1st Sess., 1985,
Serial No. 99-A. Committee Print.

U.S. Department of Health and Human Services. Bureau of Radiological Health. *An Evaluation of Radiation Emission from Video Display Terminals*, HHS Publ. No. FDA 81-8153. Washington: Government Printing Office, 1981.

U.S. Department of Health and Human Services. *Health Hazard Evaluation, United Airlines, San Francisco, California*. HETA 84-191. Cincinnati: National Institute for Occupational Safety and Health, 1985.

U.S. Department of Health and Human Services. *Health Hazard Evaluation Report, Southern Bell, Atlanta, Georgia*. HETA 83-329-1498. Cincinnati: National Institute for Occupational Safety and Health, 1984.

U.S. Department of Health and Human Services. National Centers for Disease Control. *Cluster of Spontaneous Abortions*. EPI-80-113-2. Atlanta: National Centers for Disease Control, 1981.

U.S. Department of Health and Human Services. National Institute for Occupational Safety and Health. *Job Stress Factors in Video Display Operations*, by Lawrence Schleifer and Michael Smith. Cincinnati, National Institute for Occupational Safety and Health, 1983.

U.S. Department of Health, Education and Welfare. National Institute for Occupational Safety and Health. *Industrial Hygiene Engineering and Control, Section 6: Nonionizing and Ionizing Radiation Student Manual*. Washington: Government Printing Office, 1978.

U.S. Department of State. Office of Medical Services. *Evaluation of the Health Status of Foreign Service and Other Employees from Selected Eastern European Posts (Final Report)*, by Abraham M. Lilienfeld, James Tonascia, Susan Tonascia, Charlotte H. Libauer, and George M. Cauthen. Washington: Government Printing Office, 1978.

"VDT Health Debate." *Office Systems Ergonomics Report* 5(4):18-20 (1986).

"VDT Radiation Again Linked to Abnormal Pregnancies in Mice." *VDT News* 4(4):1, 9-11 (1987).

"VDT Radiation: New Research Suggest Biological Effects Possible." *VDT News* 1(5):4-9 (1984).

"Viewpoint." *At the Centre* 9(5):21-22 (1986).

Westin, Alan F., Heather A. Schweder, Michael A. Baker, and Sheila Lehman. *The Changing Workplace: A Guide to Managing the People, Organizational and Regulatory Aspects of Office Technology*. White Plains, NY: Knowledge Industry Publications, 1985.

Wibom, Roger I., and Lars W. Carlsson. "Work at Visual Display Terminals among Office Employees: Visual Ergonomics and Lighting." In *Work with Display Units 86*, 357-367. *See* Knave 1987.

"Workers' Compensation Update." *VDT News* 2(6):3 (1985).

World Health Organization. *Radiofrequency and Microwaves: Environmental Health Criteria 16*. Geneva: WHO, 1981.

Zaret, Milton M. "Cataracts and Visual Display Units." In *Health Hazards of VDTs?* 47-60. *See* Pearce 1984.

Glossary

Accommodation (visual) — The contraction of the muscles of the eye which permits changes of focal distance.

Acuity (visual) — A measure of the eye's ability to resolve fine detail.

Asthenopia — Technical term for eyestrain.

Capsular cataract — An opacity which occurs in the front (anterior) or back (posterior) lens capsule of the eye.

Cataract — An opacity of the eye which may restrict vision: spontaneous cataracts occur within the lens itself; radiation-induced cataracts occur within the lens capsule.

Catecholamines — Chemicals in the blood, such as adrenaline, which act as hormones and neurotransmitters and affect the heart, blood vessels, and other organs, and which increase as a result of stress.

Cathode ray tube (CRT) — The glass tube used in most display terminals to produce the letters, numbers, and symbols; the characters are formed when an electron beam activates light-emitting phosphors.

Computer terminal — Keyboard (or other input device), monitor, and peripherals that are connected to a central computer.

Confounding factor — A variable which may interact with other variables to alter the outcome of an experiment; see **Hawthorne effect**, **Recall bias**.

Convergence — The ability of the eyes to coordinate their alignment through a single point of interest.

Copy — See **Hardcopy**.

Corticoids — Steroids produced by the adrenal cortex.

Critical fusion frequency (CFF) — The frequency at which a flickering object is perceived as producing a continuous image; varies widely from individual to individual.

Data entry — Transcription of letters and/or numbers from source copy to computer memory; also used to describe the job which is dedicated to this task, i.e., *data entry operator*.

Down time — Period of interruption (expected or unexpected) in the functioning of a computer system.

Electrostatic charge — Static electricity.

Epidemiology — The study of the distribution of a disease in a population.

Ergonomics — The scientific study of the relationship between people and work, especially between people and the machines they use; also called *human factors*, particularly in the United States.

Flicker — Image instability caused by the fading and refreshing of the phosphors which create the characters on a display screen; see also **Critical fusion frequency, Persistence**.

Flyback transformer—The component in a cathode ray tube VDT which causes the electron beam to move back and forth across the display screen in a regular pattern.

Frequency — The number of complete waveforms (cycles) emitted per second (of electromagnetic radiation); commonly measured in **Hertz**.

Glare — Visual discomfort and/or legibility impairment caused by excessively great variations in luminance between objects within the visual field .

Hardcopy — Printed document.

Hawthorne effect—A well-known confounding factor in which an observed positive experimental result is produced simply by the fact that the subjects are aware that an experiment is being conducted; named after the Hawthorne Western Electric plant, where the effect was first identified.

Hazard — A possible source of danger or peril.

Humidity — See **Relative humidity**.

Hypothesis — The description of a predicted relationship that if A is true, then B is also true; acceptance is subject to experimental proof.

Indirect lighting—Lighting in which no more than 10 percent of the light emissions reach the working plane directly.

Interlacing — A technique for doubling the apparent refresh rate by displaying alternate frames so that the raster scans interlace.

Lactic acid—A three-carbon acid produced when glucose is metabolized in the absence of enough oxygen to degrade the glucose to carbon dioxide and water; in humans, builds up in muscles and blood after exhaustive exercise or static work in which the blood flow is not sufficient; excess lactic acid causes muscle pain.

Luminance — The brightness of a surface, measured by the amount of light coming from the surface in a specific direction.

Monitor (noun)—See **Video display terminal**.

Monitor (verb)—To oversee or keep track of a worker's performance using the computer, i.e., keystroke counting.

Neural tube defects—Imperfect development of the fetal spinal column.

Palm rest—See **Wrist rest**.

PEMF — Pulsed electromagnetic field.

Persistence — A phosphor's ability to continue to emit light after excitation, measured in length of time.

Phosphor — A luminescent material which can emit light when excited by electrons; used to coat the inside of the cathode ray tube of a VDT.

Probability — Measure of the likelihood that an event will occur; in measures of statistical significance (of an experiment), the likelihood that chance occurrence could explain the observed results: represented in scientific text as p.

p — Probability (statistical abbreviation).

Radiation — Electromagnetic energy propagated through space as waves.

Recall bias—A distortion of experimental results caused by a subject's over-reporting or under-reporting of an event due to the passage of time, the subject's perceived personal involvement with the event being examined, and so on; see **Confounding factor**.

Refresh rate—The number of times per second (frequency expressed in Hertz) that an electron beam returns to .a point on the video screen to re-excite the phosphor and "repaint" the image; see **Frequency**.

Repetition strain injury (RSI)—Class of medical condition of the joints caused by repetitive motion, often rapid, forceful, and/or extreme motion; e.g., tendinitis, carpal tunnel syndrome.

Repetitive stress injury—*See* **Repetition strain injury**.

Response time—The amount of time which elapses between a command or request for data and the computer's response.

Role ambiguity—A lack of supervisory clarity regarding the performance expectations of a given job.

RSI — See **Repetition strain injury**.

Selection bias—A distortion of experimental results caused by choosing subjects who are already known to have some relationship to the question being examined, such as asking people in an optometrist's office to fill out a questionnaire about visual symptoms.

Source document—Printed matter (of any kind, including handwriting and typewriting) used as the basis for computer input.

Systems approach—Method of problem solving in which the relationships between all the interacting components (e.g., of an organization) are considered as a whole.

Teratogenic — Able to cause a birth defect.

Terminal — See **Video display terminal**.

Tilt and swivel —Features of a VDT monitor which enable it to be adjusted on both horizontal and vertical axes.

VDT — Video display terminal (abbreviation).

VDU — Visual display unit (abbreviation). See **Video display terminal**.

Video display terminal (VDT)—An electronic device consisting of an input device (e.g., keyboard), a monitor unit (e.g., cathode ray tube), and a connection to the central processing unit of a computer, on which information communicated to or stored in the computer is presented visually.

Visual display unit—Video display terminal (European).

Wavelength — The distance, in the direction of propagation, from point to point on successive waves (of electromagnetic radiation).

Wrist rest—A VDT workstation accessory designed to reduce pressure and discomfort caused by improper positioning of the hands and wrists.

Index